The Evolutionary Complexity of Endogenous Innovation

The Evolutionary Complexity of Endogenous Innovation

The Engines of the Creative Response

Cristiano Antonelli

Professor, Department of Economics and Statistics 'Cognetti de Martiis', University of Turin and Fellow, Collegio Carlo Alberto, Italy

Edward Elgar
PUBLISHING

Cheltenham, UK • Northampton, MA, USA

Published by
Edward Elgar Publishing Limited
The Lypiatts
15 Lansdown Road
Cheltenham
Glos GL50 2JA
UK

Edward Elgar Publishing, Inc.
William Pratt House
9 Dewey Court
Northampton
Massachusetts 01060
USA

A catalogue record for this book
is available from the British Library

Library of Congress Control Number: 2017960079

This book is available electronically in the **Elgar**online
Economics subject collection
DOI 10.4337/9781788113793

ISBN 978 1 78811 378 6 (cased)
ISBN 978 1 78811 379 3 (eBook)

Typeset by Servis Filmsetting Ltd, Stockport, Cheshire
Printed and bound in Great Britain by TJ International Ltd, Padstow, Cornwall

Contents

Acknowledgements

This book elaborates, implements, applies and tests the theoretical framework presented in Antonelli, C. (2017), *Endogenous Innovation: The Economics of an Emergent System*, Cheltenham, UK and Northampton, MA: Edward Elgar Publishing. For this purpose it uses and recombines material drawn from the following publications: Chapter 2: Antonelli, C. and Ferraris, G.L. (2011), 'Innovation as an emerging system property: An agent based model', *Journal of Artificial Societies and Social Simulation*, **14** (2); Chapter 3: Antonelli, C. and Ferraris, G. (2017), 'The Marshallian and Schumpeterian microfoundations of evolutionary complexity: An agent based simulation model', in A. Pyka and U. Cantner (eds), *Foundations of Economic Change: A Schumpeterian View on Behaviour, Interaction and Aggregate Outcomes*, Berlin, Heidelberg and New York: Springer, pp. 461–500; Chapter 4: Antonelli, C. and Colombelli, A. (2015), 'External and internal knowledge in the knowledge generation function', *Industry and Innovation*, **22**, 273–98; Chapter 5: Antonelli, C. and Fassio, C. (2016), 'The role of external knowledge(s) in the introduction of product and process innovations', *R&D Management*, **46**, 979–91; Chapter 6: Antonelli, C. and Colombelli, A. (2015), 'The knowledge cost function', *International Journal of Production Economics*, **168**, 290–302; Chapter 7: Antonelli, C. and Gehringer, A. (2016), 'The cost of knowledge and productivity dynamics. An empirical investigation on a panel of OECD countries', in A.N. Link and C. Antonelli (eds), *Strategic Alliances: Leveraging Economic Growth and Development*, London: Routledge, pp. 155–74; Chapter 8: Antonelli, C., Crespi, F. and Scellato, G. (2015), 'Productivity growth persistence: Firm strategies, size and system properties', *Small Business Economics*, **45**, 129–47; Chapter 9: Antonelli, C. and Ferraris, G. (2017), 'The creative response and the endogenous dynamics of pecuniary knowledge externalities: An agent based simulation model', *Journal of Economic Interaction and Coordination*, https://doi.org/10.1007/s11403-017-0194-3.

   I would therefore like to thank all my co-authors on the above contributions to this volume: Alessandra Colombelli, Francesco Crespi, Claudio Fassio, Gianluigi Ferraris and Giuseppe Scellato – all of the Bureau of Research on Innovation, Complexity and Knowledge (BRICK), Collegio Carlo Alberto, Moncalieri, Italy – and Agnieszka Gehringer of the Flossbach von Storch Research Institute, Cologne, Germany.

1. The engines of the creative response: the introductory framework

Cristiano Antonelli

1. INTRODUCTION

The Schumpeterian notion of creative response provides a consistent framework in which it is possible to articulate a comprehensive and coherent account of the endogenous determinants of the introduction of innovations. Firms caught in out-of-equilibrium conditions try to react to unexpected conditions of product and factor markets, and hence levels of profitability and performances away from the normal. Their reaction can be either adaptive or creative. When adaptive responses prevail, firms can only change their techniques in the existing map of isoquants: the system converges to equilibrium. When their response is creative firms can actually introduce new technologies that change the existing map of isoquants. The chances that the reaction is creative and the introduction of innovations successful are contingent upon the amount of knowledge externalities the system in which the firms are embedded is able to provide. The availability of external knowledge at costs below equilibrium levels supports their creative response and makes the introduction of productivity-increasing innovations possible.

The introduction of innovations feeds further out-of-equilibrium conditions that in turn push firms towards creative responses that may succeed, again with the eventual introduction of new innovations, provided the dynamics has not reduced the quality of knowledge governance mechanisms. In this case the system enters a positive loop of feedbacks where all the components – out-of-equilibrium conditions, knowledge generation, knowledge governance and innovation – are endogenous. The tools of evolutionary complexity apply.

When the system is not able to provide access to knowledge spillovers at low costs, the response of firms is doomed to be adaptive. When the system does not provide the necessary access at low costs to the stock of quasi-public knowledge, firms can try to change their techniques rather than their technologies: the system gravitates around equilibrium conditions without growth and change. The tools of equilibrium economics apply.

1

When the response of firms to out-of-equilibrium conditions is creative and strong, and the system supports them with persistent knowledge externalities that provide access to the stock of the existing quasi-public knowledge at low costs, the system is able to foster the rate of technological change and reproduce out-of-equilibrium conditions that may last until the quality of knowledge governance mechanisms stays put (Antonelli, 2008, 2011, 2015a, 2017a, b).

The dynamics of the innovation process is fully endogenous to the system and exhibits the typical characteristics of an emergent system property (Arthur, 2007, 2009, 2014; Foster and Metcalfe, 2012). The successful introduction of innovation is in fact the result of the interaction between individual action and the properties of the system (Roberts et al., 2017).

This chapter contributes to the framework outlined so far with analysis of the role of: i) the levels of reactivity of firms to out-of-equilibrium conditions; ii) the quality of knowledge governance mechanisms at work within economic systems that define the actual amount of knowledge externalities available to reactive firms in assessing the rate of technological change.

Section 2 explores the evolutionary complexity of the interaction between endogenous out-of-equilibrium conditions, creative response and knowledge externalities that can be elaborated from the foundations laid down by Schumpeter's essay 'The creative response in economic history'. Section 3 analyses the relationship between out-of-equilibrium conditions and firms' responses, focusing on the levels of firms' reactivity. Section 4 recalls the role of knowledge externalities in making the creative response possible and effective, focusing on the endogenous dynamics of knowledge governance mechanisms. Section 5 presents a simple model that enables us to explore the systemic and endogenous dynamics of the creative response. The conclusions summarize the results and explore their implications for both economic and policy analysis.

2. THE EVOLUTIONARY COMPLEXITY OF ENDOGENOUS INNOVATION

In 'The creative response in economic history', published in the *Journal of Economic History* in 1947, Joseph Schumpeter provides a synthesis of the alternative views about the relationship between performances and innovation he presented in 1939 and 1942 in *Business Cycles* and *Capitalism, Socialism and Democracy* (Antonelli, 2008, 2015a, 2017a, b).

In *Business Cycles* Schumpeter elaborates the view that firms are induced to introduce innovations to cope with decline of performance.

His historic analysis of innovation flows shows that the introduction of innovations peaks in the years of depression that follow the exhaustion of opportunities provided by previous gales of innovation. Firms are exposed to a decline in performance: the growth of output is weak; profitability falls below the average; and, ultimately, even actual losses emerge. The survival of firms is actually endangered. The introduction of innovations is regarded as a necessity to contrast the fall of performance below the average and possible risks of failure and exit. The generalized conditions of declining performance shared by many firms induce a collective innovation process that eventually leads to the emergence of new gales characterized by the complementarity and interoperability of a variety of new technologies. *Business Cycles* elaborates the 'failure inducement' mechanism, eventually articulated by Nelson and Winter (1982), according to which innovations are more likely to be introduced where profits and performance are below equilibrium and/or average levels.

A few years later, *Capitalism, Socialism and Democracy* provides an alternative framework where the relationship between performances and innovation is reversed. Firms that enjoy extra profits are more likely to engage in risky undertakings such as research and development (R&D) activities that are at the origin of the possible introduction of innovations. Firms with profits and performance above average and above equilibrium levels are more likely to introduce innovations also because they can fund the necessary research with internal financial resources. Profits above equilibrium reduce levels of risk aversion and liquidity constraints. *Capitalism, Socialism and Democracy* laid down the foundations of the well-known Schumpeterian hypothesis according to which rates of innovation are faster in oligopolistic markets characterized by the rivalry between large corporations with performances well above equilibrium levels.

The analysis of *Capitalism, Socialism and Democracy* contrasts that of *Business Cycles*. Although the focus and level of the analyses differ – the former elaborates at the aggregate level and focuses on the working of the system while the latter is typically microeconomic and impinges upon the theory of the firm – the relationship between performance and innovation is negative in the former and positive in the latter. The 1947 essay 'The creative response in economic history' seems to provide a synthesis: firms try to innovate when they try to cope with out-of-equilibrium conditions. In turn, out-of-equilibrium conditions take place both when performance is below and above equilibrium levels.

The intuition of 'The creative response in economic history' enables us to implement four important contributions that synthesize the different strands of literature that impinge upon the separate readership of the Schumpeterian legacy:

1. It introduces the reactivity function whereby innovation takes place as a response to out-of-equilibrium conditions that can be both negative (as in *Business Cycles*) and positive (as in *Capitalism, Socialism and Democracy*).
2. It enables us to operationalize the notion of procedural rationality.
3. It stresses the crucial role of the context in which the response takes place.
4. It provides the framework to grasp the endogenous relationship between out-of-equilibrium conditions and innovation.

Let us consider these in turn.

The Innovative Response

The introduction by Schumpeter (1947) of the reactivity function can be regarded as a major contribution to economics. It encompasses and generalizes a variety of approaches: from the induced technological change approach to the demand pull and the oligopolistic rivalry, including the very basic notion of technical change of microeconomics as well as the evolutionary approach. In basic microeconomics firms 'react' to changes in factor markets and in inputs costs, searching for new existing techniques on the 'given' map of isoquants. The notion of reactive response finds here its foundations. The notion of innovative response can be regarded as a direct extension of the reactive technical change when technological change is no longer exogenous but is regarded as the endogenous outcome of firms' conduct. In the demand pull approach, firms react to changes in demand for their products, enhancing the division of labour that enables them to introduce innovations. In the induced technological change approach, firms react to changes in input costs and innovate, changing the map of (no longer given) isoquants.

Since the seminal contribution of Dasgupta and Stiglitz (1980) corporate decision-making regarding R&D has been analysed within the framework of the typical reaction function of oligopolistic rivalry. The evolutionary approach elaborated by Nelson and Winter (1982) assumes that firms change their routines when their performance falls below average levels: the attempt to innovate is viewed as a way of coping with emerging failures. Lazonick (2007: 27) provides quite an interesting and well-synthesized framework:

> The rise of new competition poses a challenge to the innovating firm. It can seek to make an innovative response or, alternatively, it can seek to adapt on the basis of the investments that it has already made by, for example, obtaining wage and work concessions from employees, debt relief from creditors, or tax breaks or

other subsidies from the state (see Lazonick, 1993). An enterprise that chooses the adaptive response in effect shifts from being an innovating to an optimizing firm. How the enterprise responds will depend on not only the abilities and incentives of those who exercise strategic control but also the skills and efforts that can be integrated in its organization and the committed finance that, in the face of competitive challenges, can be mobilized to sustain the innovation process.

Lazonick (1993) limits the source of out-of-equilibrium conditions to the rise of new competition. In our approach, instead, changes in both factor and product markets, at the firm and the system level, do engender out-of-equilibrium conditions.

As a matter of fact the recent literature on the widespread surge of green technologies relies systematically on the notion of innovative reaction, stressing the positive role of the upsurge of oil prices, carbon taxes and environmental constraints as well as the strong increase in demand for low-emission-production processes, capital goods and final products as determinants of the creative reaction of firms pushed to introduce new energy-saving and green technologies by unexpected out-of-equilibrium conditions (Porter and van der Linde, 1995; Newell et al., 1999; Acemoglu et al., 2012; Aghion et al., 2016). Yet this literature reveals three major limits:

- It fails to elaborate explicitly an integrated notion of innovative response that is able to frame into a single and comprehensive context that includes different sources of out-of-equilibrium conditions.
- It portrays the relationship between out-of-equilibrium conditions and innovative response as automatic and deterministic, as if all firms facing unexpected changes in their product and factor markets might actually innovate.
- It assumes that the shocks to which firms react are exogenous, and is not able to grasp their endogenous determinants.

The Response as a Form of Procedural Rationality

Schumpeterian decision-making is far from Olympian rationality. Firms make plans on the basis of their limited knowledge of the actual and expected conditions of product and factor markets. When their expectations fail, they try to elaborate responses that are highly contextual and constrained by sunk costs. The response is a form of procedural rationality. The Schumpeterian notion of creative/adaptive response complements and enriches Herbert Simon's analysis of the intrinsic limits of knowledge and the role of bounded and procedural rationality (Simon, 1947, 1979, 1982).

The Role of Externalities

The outcome of the Schumpeterian response is not deterministic but strictly conditional on the availability of knowledge externalities. The response of firms may be creative, and leads to the actual introduction of innovations only if and when substantial knowledge externalities are available. When the quality of knowledge governance mechanisms and stocks of quasi-public knowledge are low, the actual provision of knowledge externalities falls below critical values, the reactive attempt of firms fails to be innovative and the response is just adaptive: technical change replaces technological change.

The response of firms to out-of-equilibrium changes in both their product and factor markets consists in mobilizing research activities. Such activities are necessary both to search for existing techniques that fit better with the changed conditions of product and factor markets and to introduce new technologies – that is, techniques that do not exist and do not belong to the existing map of isoquants. The search for and identification of existing techniques, new to the firm but already known, are not free and entail specific costs. In appropriate conditions determined by the properties of the system that provide substantial knowledge externalities and hence access to the stock of quasi-public knowledge at low costs, research activities enable the generation of additional knowledge that may eventually lead to the introduction of new technologies. The differences in outcome – whether it is just adaptive, as in the identification of new viable techniques among the many already available on the existing map of isoquants, or actually creative so as to enable the introduction of new technologies that reshape the map of isoquants – are determined by the amount of knowledge externalities available in the system, and hence by the bottom-line access and use costs of external knowledge (Antonelli, 2017a, b).

The Endogenous Relationship between Out-of-Equilibrium Conditions and Innovation

The introduction of innovation as the outcome of the creative response of firms to out-of-equilibrium conditions, contingent upon the quality of knowledge externalities available in the system, is itself the cause of further out-of-equilibrium conditions. Out-of-equilibrium conditions are not the result of exogenous shocks, but the endogenous consequence of the innovative response of firms. Not only are out-of-equilibrium conditions endogenous to the system, but the quality of knowledge externalities is also determined within the system and may increase as well as decrease.

The response of firms to out-of-equilibrium conditions, in fact, consists in the generation of additional technological knowledge that is necessary to introduce innovations. The additional knowledge spills into the system and affects the size and the quality of the stock of quasi-public knowledge available for the generation of new technological knowledge. At the same time the levels of access costs to the stock of quasi-public knowledge are determined by the systems of knowledge interactions and transactions between firms and other knowledge-intensive agents.

3. LEVELS OF REACTIVITY AND RESEARCH EFFORTS

It seems now useful to make a step forward in analysing the crucial role of the levels of reactivity of firms that try to cope with out-of-equilibrium conditions. Decision-making is based upon procedural rationality: on one hand firms do not command the understanding of all the possible alternatives; on the other they are able to explore uncharted waters and introduce innovations and change their routines. At each point in time they try to cope with the changing conditions of product and factor markets under constraints of sunk costs and past decisions. Their ability to cope with changing economic conditions is the outcome of a variety of factors, including: the type of structure and organization; the role of shareholders and stakeholders; industrial relations and levels of entrepreneurship of top managers. Figure 1.1 compares research effort (R) with performance levels both above and below the average (Π). It exhibits two levels of reactivity. Around equilibrium levels, at the intersection of the two axes, firms do not try to change their production processes. Liquidity constraints and risk aversion restrain research efforts. The further away from equilibrium, the stronger is the likelihood that firms try to change their production

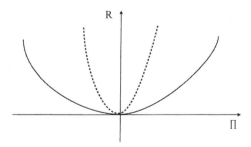

Figure 1.1   Performance and research effort: levels of reactivity

processes either by searching for new existing techniques or by actual research activities that, provided substantial knowledge externalities are available, may enable the introduction of new technologies.

Two mechanisms are at work in this process: the failure inducement articulated by Schumpeter in *Business Cycles* and the success inducement analysed in *Capitalism, Socialism and Democracy*. Let us consider them in turn. Firms try to innovate to cope with the high risks of failure in the right quadrant where losses are greater, performance is worst and research efforts stronger. The negative conditions of performance and the high risks of failure reduce the risk aversion. The substitution of tangible investments with intangible ones and the increase in research activities are the last chance to try to cope with threats to survival. In the left quadrant, instead, the larger the profits, the better the performance and the stronger the research efforts. Firms can fund a larger budget of research activities that may put them in a position to try to innovate because of an abundance of internal cash. Large internal cash reserves reduce liquidity constraints, avoid the credit rationing of external finance and reduce the levels of risk. The possible failure of innovative undertaking does not put at risk the entrepreneurial managers who have already secured high levels of profitability for their shareholders and stakeholders. The success of the risky undertakings may yield further growth and larger profits that would benefit the managers.

Relationships between out-of-equilibrium conditions, however, can take place with different levels of elasticity. Figure 1.1 exhibits two levels of reactivity. The bold line represents low levels of reactivity: firms are reluctant to change their levels of innovative efforts. The dotted line represents high levels of reactivity.[1]

The key point is that the extent to which the change in the levels of (relative) performance affects the levels of research efforts. The bold line represents high levels of reactivity stemming from high levels of entrepreneurship. Firms guided by managers with high levels of entrepreneurship are more likely to exhibit high levels of reactivity to changing levels of profitability and performance at large. Firms guided by managers with low levels of entrepreneurship are less reactive.

Figure 1.2 compares levels of reactivity with levels of research effort: on the horizontal axis $\Delta\Pi$ measures *in absolute terms* the differences between profitability and performance of each firm and the normal and/or average profitability and performance of all the firms in the system; on the vertical axis R measures levels of research effort. The levels of reactivity play an important role in assessing the amount of research carried out in order to cope with out-of-equilibrium conditions. Large research budgets may implement a creative response and introduce innovations

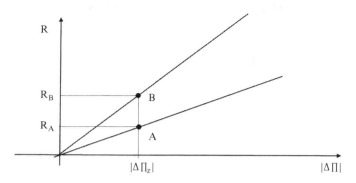

Figure 1.2   Levels of reactivity and research effort

when large knowledge externalities that reduce access costs to the stock of quasi-public knowledge are made possible by the quality of knowledge governance mechanisms at work within the system.

Figure 1.2 shows that at the same level of out-of-equilibrium conditions measured by the difference in absolute terms between profitability and performance and the normal levels of profitability and performance ($\Delta\Pi_z$) there are two quite different levels of research efforts: R_A with low levels of reactivity and R_B with high levels of reactivity. The levels of reactivity play a major role in assessing the elasticity of the system to out-of-equilibrium conditions.

The implications of this analysis are important to assess the actual determinants of the innovative efforts of an economic system. The amount of research is likely to be larger in systems characterized not only by widespread out-of-equilibrium conditions than in systems where all firms operate near equilibrium and the variance of profitability and performance at large is small, but also by high rather than low levels of reactivity.

These implications can be amplified when we measure out-of-equilibrium conditions in terms of variance with respect to average profitability and performance rather than with respect to normal profitability and performance. When average profitability and performance is taken into account as the relevant measure of out-of-equilibrium conditions we expect that in systems with average profitability and performance above normal levels but low levels of variance the reactivity is always lower than in systems where out-of-equilibrium conditions are measured with respect to normal profitability and performance. The greater the variance with respect to the average, the larger the chances that firms try to react. When the out-of-equilibrium conditions of each firm refer to average profitability and performance it becomes immediately clear that the larger is the variety and heterogeneity of firms, the larger the innovative efforts. With greater research efforts, and given levels of knowledge externalities, there are greater chances of faster rates of introduction of innovation and increase of total factor productivity and ultimately economic growth.

The results of the replicator analysis, according to which the greater the variety of firms the larger the growth rates, are confirmed, overcoming the Darwinistic and exogenous assumptions of the replicator analysis (Metcalfe, 1998).

The replicator analysis, in fact, assumes the heterogeneity of firms in terms of given and exogenous, or randomly determined, differences in fitness among species competing for scarce resources in a given environment. The larger the variance in terms of levels of fitness, the larger the rates of growth simply because the eventual survival of the species with higher fitness parameters and the exit of the weaker ones leads to a larger population and faster rates of growth along the substitution process.

In the approach outlined so far, instead, the larger is the variety and heterogeneity of firms, the larger are the research efforts that may eventually lead, with high levels of knowledge externalities provided by the system, to a faster rate of introduction (and creative adoption) of technological innovations, and hence larger rates of increase in total factor productivity (TFP). The positive relationship between heterogeneity, variety and variance of profitability, performances and rates of growth is confirmed. The determinants of the relationship, however, are completely different. In the standard replicator analysis, innovation is exogenous. Growth is determined by the diffusion of the exogenous innovation. In the analysis implemented so far, innovation is endogenous.

4. LEVELS OF KNOWLEDGE GOVERNANCE AND THE CREATIVE RESPONSE

The creative response of firms to out-of-equilibrium conditions is contingent upon the actual costs of knowledge. Knowledge costs are determined by the knowledge externalities available in the system. Knowledge externalities are pecuniary and diachronic. They make possible the use of knowledge spillovers at costs that are below the equilibrium levels of knowledge as a standard good. In turn the amount of knowledge externalities available in a system depends upon the quality of knowledge governance mechanisms at work in a system (Antonelli, 2017a, b).

Knowledge governance consists in the structure of knowledge generation activities; the organization of the architecture of knowledge of interactions and transactions; and the institutional set-up that makes the accumulation of the stock of quasi-public knowledge possible and enables the use of that knowledge at low costs (Ostrom and Hess, 2006).

Because of the limited appropriability of knowledge, inventors can retain full control of the economic benefits stemming from the new technological knowledge they have generated only for a limited stretch of time. After that appropriation window, technological knowledge becomes a quasi-public good and contributes to the accumulation of a stock of quasi-public knowledge that third parties can try to access as an indispensable complementary input in the recombinant generation of further knowledge (Weitzman, 1996).

High-quality knowledge governance mechanisms favour knowledge interactions along the vertical stages of the inter-sectoral division of labour with effective user–producer interactions that also include final markets, effective knowledge transmission between public and private research centres. They also reduce the exclusivity of intellectual property (IP) rights so as to support both the necessary rewards of knowledge producers and the widespread secondary uses of proprietary knowledge for the recombinant generation of new knowledge (Antonelli, 2015b).

The quality of knowledge governance mechanisms at work in the system plays a central role in this process on two counts: first, accumulation of the knowledge spilling from 'inventors' into the stock of quasi-public knowledge is contingent upon the quality of the knowledge governance mechanisms. In systems with poor knowledge governance mechanisms the uncontrolled spillover of knowledge dissipates in the atmosphere and results in slow rates of accumulation of the stock of quasi-public knowledge. In contrast, knowledge spillover effectively adds to the existing stock of quasi-public knowledge in systems endowed with high-quality knowledge governance mechanisms and low levels of dissipation. Second, access to and use of knowledge spilling from third parties accumulated in the stock of quasi-public knowledge is not free. Relevant absorption costs are necessary in order to search, identify, decode, access and finally (re)use the knowledge available in the system. Knowledge absorption costs are reduced by effective knowledge governance mechanisms that favour search, screening and access to existing knowledge for new uses.

The actual levels of knowledge externalities, and hence of the costs of external knowledge, are determined by the amount of quasi-public knowledge and the amount of absorption activities that are necessary to benefit from and use it. When high-quality knowledge governance mechanisms are at work, firms can access external knowledge at low costs, far below equilibrium levels, both because of low absorption costs and the large amount of quasi-public knowledge. Productivity-enhancing innovations depend upon the actual access to knowledge spillovers that make external knowledge indispensable in the recombinant generation of new knowledge, available at costs that are below equilibrium levels. In these systems, consequently, firms that try to cope with out-of-equilibrium conditions have more chances to implement creative responses and introduce technological innovations that reshape the map of isoquants.

Firms embedded in systems with poor knowledge governance mechanisms experience high absorption costs of knowledge spillovers. The final costs of external knowledge are larger, actually close to the levels of knowledge costs if it were a standard economic good. These firms cannot take advantage of knowledge externalities. Their response to emerging out-of-equilibrium conditions is consequently adaptive. They try to cope with out-of-equilibrium conditions by means of technical changes that enable them to move on the existing map of isoquants.

The introduction of productivity-increasing innovations is strictly contingent upon the properties of the system. For given levels of reactivity, the response of firms is actually creative according to the amount of knowledge externalities available in the system and their success – in terms of actual introduction of productivity enhancing innovations – is ultimately determined by: i) the actual costs of the external knowledge that is an indispensable input strictly complementary to the research efforts in the recombinant generation of new knowledge; and ii) the actual bottom-line costs of the knowledge that enters the technology production function as a complementary input alongside the traditional tangible ones such as capital and labour.

Hence, for given levels of reactivity, a system and each agent within the system have more chances to select a creative rather than an adaptive response to out-of-equilibrium conditions, according to costs of access to the stock of quasi-public knowledge that, in turn, depend upon the quality of knowledge governance mechanisms at work in the system.

This result complements the outcome of the previous section according to which, for given levels of knowledge externalities, the larger the levels of reactivity, the larger the research efforts and hence the rate of introduction of innovations.

The analysis also makes it clear that the quality of knowledge governance mechanisms is fully endogenous: it is continually shaped and reshaped by firms' conduct, by their levels of reactivity and by the actual rates of generation of new technological knowledge and eventual introduction of innovations. The quality of knowledge governance mechanisms may stay put through time, as well as improve and deteriorate. These processes are typically non-ergodic and yet far from deterministic: typically path rather than past dependent.

5. THE DYNAMICS OF THE CREATIVE RESPONSE: A GRAPHIC EXPOSITION

The elements introduced so far to explore the role of the engines of creative response can now be nested into a fully fledged system of interdependence that relates creative response to levels of reactivity, and hence to the actual amount of research that takes place in a system, the consequent amount of technological knowledge that can be generated taking into account levels of knowledge externalities and, consequently, the extent to which the response is creative and its effects in terms of the rate of introduction of innovations and the amount of output and TFP that can be achieved (see the Appendix).

Quadrant I of Figure 1.3 (starting from northwest) reproduces simply Figure 1.2. Starting with a given level of $\Delta\Pi$, the margin of actual profitability and performances with respect to normal (average) profitability, this quadrant shows the effects of different levels of reactivity. With high levels of reactivity, firms are induced to the innovative effort R_B clearly larger than R_A that would take place with low levels of reactivity.

Quadrant II represents knowledge generation activities.[2] Knowledge generation activities are far more productive when they can rely upon strong knowledge externalities that enable firms to access the stock of quasi-public knowledge at low costs. The innovative efforts yield a larger knowledge output (T), respectively found in F and G for more and less reactive firms. When knowledge externalities do not support the generation of technological knowledge and access costs to the stock of the quasi-public

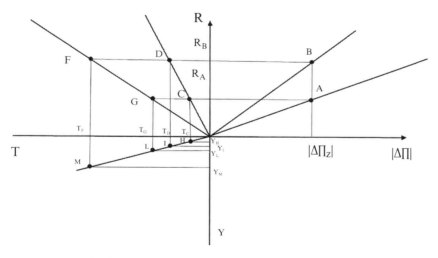

Figure 1.3   The dynamics of the creative response

knowledge are close to the equilibrium levels of knowledge – that is, as if it were a standard economic good with high levels of exhaustibility and appropriability[3] – output is lower: respectively D and C for more and less reactive firms.

Quadrant III shows production activities. Technological knowledge (T) enters the production function as an input alongside the traditional tangible inputs, capital and labour. Assuming fixed levels of capital and labour, this quadrant shows the effects of the larger amount of technological knowledge generated in Quadrant II on the production of output (Y). For given levels of reactivity, when T is larger than in equilibrium because of the positive effects of knowledge externalities, Y is larger than it would be when T matches equilibrium levels. As Quadrant III shows, when reactivity levels are high (R_B) and knowledge externalities are effective, knowledge output T_F is larger than T_D at the levels that take place when knowledge externalities do not support the generation of knowledge at costs that are below equilibrium levels. By the same token, when reactivity levels are low (R_A) and knowledge externalities are effective, knowledge output T_G is larger than T_C at the levels that take place when knowledge externalities do not support the generation of knowledge at costs that are below equilibrium levels. The distances on the inferior vertical axis – (T_G T_C) and (T_F T_D) – measure the effects of the lower costs of technological knowledge on output Y and, as such, provide a reliable clue to the effects of knowledge externalities on total factor productivity.

It is clear that a system endowed with high levels of both the reactivity of firms and the quality of knowledge governance mechanisms is better able to increase not only its rates of growth but also, and most importantly, its rates of increase of TFP: (T_G T_C) > (T_F T_D).

Firms' levels of reactivity to out-of-equilibrium conditions and the quality of knowledge governance mechanisms that defines the actual amount of knowledge externalities are the drivers of the creative response. The larger the reactivity of firms and the higher the quality of the knowledge governance mechanisms, the larger the rates of introduction of innovations, as measured by the amount of technological knowledge generated, and the growth of both output and TFP.

The system of interdependencies framed in Figure 1.1 provides the building blocks to explore Quadrant IV and to study the dynamics of the system. It is in fact clear that:

● The larger the variance of profitability and performances levels, the higher the reactivity levels and the larger the amount of research efforts.

- The larger the amount of research efforts and the lower the costs of accessing and using the stock of quasi-public knowledge, the lower is the actual output of the knowledge generation activities. With a given budget firms that enjoy relevant knowledge externalities can take advantage of costs to access the stock of quasi-public knowledge that are below equilibrium levels and can generate more knowledge at costs that are below equilibrium.
- The larger the knowledge output, the lower its costs and the larger the output Y of the technology production function and the levels of TFP.
- The dynamics of the system *for given and invariant levels of the quality of knowledge governance mechanisms* exhibits the typical traits of a self-sustained process supported by positive feedback.
- Because of the working of diachronic knowledge externalities according to which the flows of proprietary knowledge add to the stock of quasi-public knowledge, after a limited window of appropriation the larger the knowledge output at each point in time and, for invariant levels of knowledge governance, the larger the stock of quasi-public knowledge. Hence the lower are its costs and, consequently, knowledge costs are lower and the chances that firms are able to implement an effective creative response are higher.
- The faster innovations are introduced and the larger the growth of TFP, the larger are likely to be the unexpected changes in factor and product markets that are at the origin of out-of-equilibrium conditions levels of variety within the system. Variety and heterogeneity are more likely to be persistent, and actually may amplify in systems characterized by fast introduction of innovation. A virtuous self-feeding process of creative disorder can actually enter in place, provided the quality of knowledge governance mechanisms is also persistent.
- The lower the rates of introduction of innovation, due to low-quality knowledge governance mechanisms, and hence slow accumulation of the stock of quasi-public knowledge and high costs to access it, the lower the chances that creative responses actually can take place. The prevalence of adaptive responses reduces the heterogeneity of firms. The typical Marshallian search for equilibrium displays fully its effects: worst-performing firms are sorted out and the benchmark quality of outstanding ones is rapidly imitated by all the other firms. At the end of the Marshallian process heterogeneity is substituted by homogeneity, and equilibrium conditions prevail.

The quality of knowledge governance mechanisms is fragile and fully exposed to events that may take place along the process and change it. The generation of new technological knowledge and the introduction of new technologies, at each point in time, can change (sometimes radically) the structure of the system, the organization of knowledge interactions and transactions, the architecture of knowledge networks and the institutional setting that qualify the knowledge governance mechanisms. Consequently, the system dynamics is path dependent, as opposed to past dependent, as it exhibits strong historic, non-ergodic elements, such as the amount of quasi-public knowledge, that depend upon the accumulation of generation flows. At the same time it is also exposed to possible degeneration of the quality of knowledge governance mechanisms brought about by the very dynamics of the process.

The decline in quality of knowledge governance mechanisms can easily stop the system dynamics, with two distinct negative effects: first, it reduces the rates of accumulation of the flows of new technological knowledge, and hence the increase in the stock of quasi-public knowledge. This has direct negative effects on the rates of reduction of the access costs of external knowledge that firms use as a necessary complementary input in the recombinant generation of new technological knowledge. Second, it increases the levels of absorption costs, and thus may actually lead to an increase in the cost of external knowledge. The consequences are clear. The likelihood that firms can implement creative, rather than adaptive, responses declines with the consequent reduction in rates of technological change and increase in total factor productivity.

6.   CONCLUSIONS

The notion of creative response is at the same time the synthesis of the Schumpeterian legacy and the foundation stone of a comprehensive platform that uses the tools of evolutionary complexity able to accommodate in a coherent framework the understanding of endogenous innovation as an emergent system property. The notion of creative response enables us to go beyond the limitations and shortcomings of the evolutionary approaches that build upon biological metaphors. Innovation can be analysed as the outcome of the interdependence between individual decision-making and the properties of the system in which it takes place. The outcome of individual decision-making – the actual introduction of innovations – depends upon the characteristics of the system. The latter in turn is influenced by the outcomes of individual decision-making.

This chapter has explored the engines of the creative response: the levels of firms' reactivity to out-of-equilibrium conditions and the properties of the system that support the creative response with the provision of knowledge externalities that enable innovating firms to access and use the stock of quasi-public knowledge to generate new knowledge at costs that are below equilibrium levels.

The analysis has shown that low levels of reactivity reduce the amount of innovative efforts a system is able to mobilize. At the same time, high levels of reactivity without the support of appropriate levels of knowledge governance favour the rapid return to equilibrium levels, but not the actual introduction of innovations. High-quality knowledge governance mechanisms coupled with low levels of reactivity lead to slow rates of introduction of innovations and slow rates of increase in output and total factor productivity. In contrast, a system characterized by high levels of reactivity and high levels of knowledge governance enables firms to implement creative responses that lead to rapid introduction of innovation and fast rates of increase in output and TFP.

Systems characterized by high-quality and consistent knowledge governance mechanisms and reactive managerial styles are likely to experience rapid introduction of innovations that feeds persistent growth via: i) the recreation of out-of-equilibrium conditions; ii) the accumulation of the stock of quasi-public knowledge; iii) the quality of knowledge governance mechanisms; and iv) the persistence of knowledge externalities. In such systems creative disorder is persistent and may last as long as the quality of knowledge governance mechanisms is able to cope with the dynamics of the system and is fortified rather than damaged by it. An endogenous loop of positive feedback supports the growth of the system and the persistence of out-of-equilibrium conditions.

Systems characterized by low-quality knowledge governance mechanisms and managerial styles with low levels of reactivity are doomed to converge rapidly to equilibrium. The Marshallian search for equilibrium prevails when adaptive responses prevail over creative ones. The adaptive response prevails when the quality of knowledge governance mechanisms is poor, and, consequently, the amount of knowledge generated at each point in time is small and cumulates slowly, the access and secondary use costs of the stock of quasi-public knowledge are high, and search activities enable firms to engage in technical rather than technological changes. The variance of firms is quickly reduced by the exit of the worst-performing ones and the imitation of the advanced ones. Variety decreases together with variance and the slowdown of the rates of innovation, of increase in total factor productivity and in growth of output.

The implications for economic policy are clear. First, a managerial style able to integrate high levels of entrepreneurship helps improve system performance. The dichotomy between entrepreneurs and managers, where the former are associated with small firms and start-ups and the latter with incumbent corporations, should be abandoned. Creative managers of large corporations are just as necessary as competent entrepreneurs in small firms and newcomers. Second, the quality of the knowledge governance mechanisms that rule the accumulation of the stock of quasi-public knowledge and its access and secondary use at low costs is a central asset of an economic system that assigns a central role to the rate of introduction of technological and organizational innovations.

Public policy should care about: the architecture of inter-sectoral flows of knowledge along the multi-stage user–producer interactions; the quality of the public research infrastructure; the interface between public and private research centres; the mobility of skilled personnel among and between firms and the public research system; the working of the knowledge markets; the role of knowledge-intensive business services (KIBS); the exclusivity of intellectual property rights so as to favour at the same time the appropriation of economic benefits stemming from the introduction of innovations as well as the useful dissemination and secondary uses of existing technological knowledge.

NOTES

1. In Figure 1.1 the quadratic relationship is symmetric: the response of firms to performances above and below the average is specular in shape. Internal financial constraints and credit rationing might reduce the ability of firms with below average performance – and even more below normal levels of profitability – to fund the necessary research activities. At the same time, however, firms performing above the average may use part of their profits to pay higher dividends to shareholders and higher wages to employees, and to fund other stakeholder benefits, reducing the amount of resources that can be used to fund internal R&D. The actual shape of the quadratic relationship is determined by the institutional set-up of financial markets, intellectual property rights, industrial relations, and product and factor markets.
2. The geometric representation posits contant returns to scale in knowledge generation activities. Diminishing returns to scale might be easily accommodated with a negative second derivative without altering the basic relationship.
3. Arrow (1962) identifies the special features of knowledge such as limited appropriability and exhaustibility, substantial indivisibility, cumulability and complementarity, and low costs of reproduction by confronting knowledge with respect to standard economic goods. The negative effects of the limited appropriability of knowledge on the incentives to its generation, with high-quality knowledge governance mechanisms and substantial levels of knowledge cumulability and indivisibility, can be more than compensated for by their positive effects in terms of spillovers and the consequent reduction in costs of knowledge below equilibrium levels.

APPENDIX: A SIMPLE MODEL

Following the literature that impinges upon the constructionist design methodology (CDM) approach (Crépon et al., 1998), analysis of the engines of the creative response can be framed in a simple system of equations: i) the research function; ii) the knowledge generation function and its cost equation; iii) the external knowledge cost equation; and iv) the technology production function. Let us introduce them in turn.

The research function summarizes the relationship between out-of-equilibrium conditions as measured by $\Delta\Pi$, that is, the difference between the levels of profitability and performance of each firm and normal ones in absolute terms – in other words, taking into account both profit and performance above and below the norm:

$$R = f (j \ \Delta\Pi) \tag{1A.1}$$

where j measures the levels of reactivity.

The knowledge generation function formalizes the relationship between research efforts (R) and the actual output of knowledge (T) taking into account the stock of quasi-public knowledge (SQPT) available in the system, where m and n are their output elasticity. The Cobb–Douglas specification of the knowledge generation function makes explicit the strict complementarity between the stock of external knowledge drawn from the stock of quasi-public knowledge available in the system and the flow of internal research efforts. The cost equation includes on the left-hand side the amount of the research budget (R) that has been determined by equation (1A.1) and, on the right-hand side, the unit costs (r) of (R&D) activities and the search costs (u) that enable firms to access and use the stock of quasi-public knowledge:

$$T = h (R^m \ SQPA^n) \tag{1A.2}$$

$$R = rR\&D + uSPQT \tag{1A.3}$$

The size of the stock of quasi-public knowledge is fully endogenous. Because of diachronic knowledge externalities, in fact, it depends on the amount of knowledge flows that have been generated in previous time periods and the quality of knowledge governance mechanisms (KGM) that rule their accumulation process:

$$SQPT = l\left(KGM, \left(\sum_{t=n}^{N} T_t \right) \right) \tag{1A.4}$$

The costs of accessing and using the stock of quasi-public knowledge are also endogenous as they depend on its size (SSPQT) and KGM:

$$u = m \text{ (SSPQT, KGM), where } h' < 0 \qquad (1A.5)$$

The unit cost of technological knowledge (z) is now fully endogenous:

$$z = R/T \qquad (1A.6)$$

Finally, the technology production function specifies the relationship between output (Y), the standard inputs capital (K) and labour (L) and knowledge (T) produced in the upstream knowledge generation function, with their respective output elasticity α, β and γ. Next to it is the standard cost equation, where r measures capital user costs, w wages and z the actual level of the cost of knowledge generated upstream that takes into account the effects of knowledge externalities. Equation (1A.7) includes the measure (A) of total factor productivity:

$$Y = A \, (K^{\alpha} \, L^{\beta} \, T^{\gamma}) \qquad (1A.7)$$

$$C = rK + wl + zT \qquad (1A.8)$$

Because zT = R, it is evident that, for endogenous levels of R the lower the endogenous levels of z, the larger is T. Hence the levels of TFP are determined by the difference between the equilibrium levels of the cost of knowledge (g) that would take place if it were a standard economic good and the actual costs of knowledge (z) that take into account the effects of upstream knowledge externalities:

$$A = n \, (g - z) \qquad (1A.9)$$

When z = g firms are not able to introduce productivity-enhancing innovations. The introduction of such innovations takes place only when g > z, when the generation of technological knowledge can rely on effective knowledge externalities that reduce the cost of external knowledge (u) below equilibrium levels so that also the costs of technological knowledge (z) as an intermediary and yet indispensable input in the technology production function are below equilibrium levels g (Antonelli, 2013a, 2017a, b).

2. Innovation as an emergent system property

Cristiano Antonelli and Gianluigi Ferraris

INTRODUCTION

The chapter develops an agent-based simulation model (ABM) of innovation considered as an emerging property of a complex system. It explores how architectural, organizational and institutional variables – such as the spatial distribution of firms and intellectual property rights regimes – have an impact on innovative behaviours. Firms are considered as myopic agents that may react creatively to unexpected events. Their reaction may be adaptive or creative, according to the localized context of action. The reaction of agents may lead to the introduction of productivity-enhancing innovations if and when the organization of the system is such that the reactive agents can actually take advantage of external knowledge available within the innovation system in which they are embedded. In this approach external knowledge is an indispensable factor, together with internal research activities, in the generation of new knowledge.

Our approach contributes a line of enquiry of evolutionary economics that emphasizes the role of interactions among agents within the organized complexity of economic systems. This approach differs from evolutionary analyses of Darwinistic ascent where innovation is spontaneous and occurs randomly, in-house capacities are considered unique sources of novelty-creating activities and markets are credited with the role of selecting alternative novelties (Penrose, 1959; Nelson and Winter, 1982).

In our approach innovation is an emerging property at the system level that takes place when the actions of individuals and the organization of the system match. Knowledge interactions among heterogeneous agents and the organization of the knowledge flows within the system play crucial roles in assessing the chances of individual firms actually introducing innovations. Access to external knowledge, together with internal learning and research, is viewed as indispensable for the generation of new knowledge. The introduction of innovations is analysed as the result of systemic knowledge interactions between myopic agents that are credited with an extended procedural rationality that includes forms of reaction. Such reactivity can be either adaptive or creative. The reaction of agents can be creative so as to engender the introduction of productivity-enhancing innovations when a number of contextual conditions that enable access to external knowledge are fulfilled (Anderson et al., 1988; Lane et al., 2009; Zhang, 2003; Antonelli, 2011).

The aim of the chapter is to show that, because of the relevance of external knowledge for the generation of new knowledge, the organization of the system articulated in the different institutional and architectural settings of the structure in which knowledge interactions take place affects the rate of new knowledge generation and the introduction of technological innovations (Bischi et al., 2003). Using ABM methodology, the chapter shows that innovation is likely to emerge faster and better in organized complex systems characterized by high levels of dissemination and accessibility to knowledge externalities.

The rest of the chapter is structured as follows. The following section elaborates the theoretical framework and presents the building blocks of an approach that integrates the economics of innovation and complexity. We then present the ABM of the innovation system and the results of the simulation, focusing on an alternative hypothesis about the institutional and architectural features of the innovation system. The penultimate section elaborates the policy implications of the results before the conclusions summarize the main results and put them into perspective.

THE THEORETICAL FRAMEWORK

This section presents the basic assumptions and hypotheses about an economic system where innovation is characterized as the emergent property of the system dynamics of knowledge interactions. The introduction of innovations is analysed as the possible result of systemic interactions between heterogeneous and myopic yet learning and reactive agents when and if they can take advantage of external knowledge to make their reactions creative as opposed to adaptive.

A Behavioural Approach Enriched by Creativity

There are direct links between the Schumpeterian legacy and the behavioural theory of firms that have been poorly appreciated so far. Schumpeter (1947) in a landmark contribution introduces the notion of creative reaction as a conclusive point of his theoretical elaboration. Schumpeterian firms are portrayed explicitly as myopic agents that are unable to foresee all possible events and are occasionally surprised by unexpected events. Schumpeterian firms are myopic but have the ability to react and to rely on external resources in their reaction. Their reaction to changing economic environments can be either adaptive or creative. If their reaction is adaptive, equilibrium conditions prevail and lead to traditional price/quantity adjustments with no innovation. Their reaction becomes creative, as opposed to adaptive, when knowledge interactions supported by a viable organization of the system makes it possible to access external knowledge in favourable conditions.[1] Creative reaction engenders out-of-equilibrium conditions and, with appropriate external conditions, feeds a virtuous cycle of growth and change (Antonelli, 2011).

This Schumpeterian legacy is fully consistent and actually complementary with the basic assumptions of organizational behavioural theory elaborated by Jamie March and Herbert Simon (1958). In classic behavioural theory firms are myopic: their rationality is bounded, as opposed to Olympian, because of the wide array of unexpected events, surprises and mistakes that characterize their decision-making and business conduct in an ever-changing environment. Firms, however, are endowed with an extended procedural rationality that includes the capacity to learn. Agents are intrinsically heterogeneous. They have distinctive and specific characteristics that qualify their competence, the endowment of tangible and intangible inputs and their location in the space of interactions (Cyert and March, 1963; March, 1988, 1991).

In our approach agents can do more than adjust prices to quantities, and vice versa: they can try to react to changing conditions in their economic environment by means of the generation and exploitation of technological knowledge through the introduction of technological innovations. To innovate, firms mobilize their slack resources consisting in tacit knowledge and competence accumulated by means of internal learning processes (Leibenstein, 1976). Internal slack resources, however, are a necessary but not sufficient condition to innovate. Reaction becomes creative only with the support of an organized complexity of the system in which firms are embedded.

A behavioural theory of a myopic but learning firm enriched by Schumpeterian creativity provides the basis for implementing a model of the economic complexity of technological change. In our approach firms try to innovate when their performance differs sharply from the average. Clear causality between performance, both negative and positive, is established. When performance is below average firms are dissatisfied and try to change their routines. When performance is above average firms have more opportunities to fund risky activities. This out-of-equilibrium causal link, in a typical satisficing approach between performance and attempts to innovate, marks a clear difference from the approach post-Nelson and Winter where no causality is introduced and innovation is viewed as the spontaneous result of the behaviour of firms considered as individual agents.

Innovation and Knowledge

The introduction of technological and organizational innovations requires the generation of new knowledge. The generation of knowledge is characterized by specific attributes: knowledge is simultaneously the output of a specific activity and an essential input in the generation of new knowledge. Because of knowledge indivisibility, access to existing knowledge at any point in time is a necessary condition for the generation of new knowledge. No firm can command all available knowledge; hence no firm can generate new technological knowledge alone. The twin character of knowledge as output of research and input in the generation of further knowledge stresses the basic complementarity and interdependence of agents in the innovation process: innovation is inherently the collective result of the interdependent and interactive intentional action of economic agents (Blume and Durlauf, 2001, 2005).

The structure of the system and its continual change, following Simon Kuznets's analysis, play crucial roles. The system's organization and structure affect the architecture of knowledge externalities, interactions and transactions and play crucial roles in access to external knowledge, and hence in defining the actual chances of agents implementing their reactions and making them creative as opposed to adaptive (Silva and Teixeira, 2009).

Technological knowledge is viewed as the product of recombining existing ideas, both diachronically and synchronically. The generation of new knowledge stems from the search and identification of elements of knowledge that had not been considered previously, and their subsequent active inclusion and integration in the pre-existing components of each firm's knowledge base (Weitzman, 1996, 1998; Fleming and Sorenson, 2001).

Marshallian externalities as implemented by the notion of generative interactions play a central role in this approach (Lane and Maxfield, 1997). The amount of knowledge externalities and interactions available to each firm influences their ability to generate new technological knowledge, hence the actual possibility to make their reaction adaptive rather than creative and able to introduce localized technological changes. Each myopic agent has access only to local knowledge interactions and externalities; that is, no one agent knows what every other agent in the system at large knows. Because of the localized character of knowledge externalities and interaction, location in a multidimensional space, in terms of distance between agents and their density, matters. Interactions in fact are localized as opposed to global. At each point in time agents are rooted within networks of transactions and interactions that are specific subsets of the broader array of knowledge externalities, interactions and transactions that take place in the system. In the long term, however, they can move in space and change their location in the networks. In so doing they change the organization of the system.

Contingent Factors Influencing Innovative vs. Adaptive Behaviours

Appropriate structural and institutional characteristics of the system upgrade the reaction of firms and help make it actually creative, and hence engender the introduction of productivity-enhancing innovations. Only when the role of such external and complementary systemic conditions is taken into account can the role of innovation as the productivity-enhancing result of an intentional action be articulated. System organization plays a key role as it shapes access to external knowledge. When the role of the external context is properly appreciated, it becomes clear that innovation is not only the result of the intentional action of each individual agent; it is also the endogenous product of system dynamics. Individual action and the organization of the system conditions are crucial and complementary ingredients in explaining the emergence of innovations (Lane et al., 2009).

Positive feedback takes place when the external conditions in which each firm is localized qualify the access to external knowledge so as to make the reaction of firms creative as opposed to adaptive. When the access conditions to local pools of knowledge enable the actual generation of new technological knowledge and feed the introduction of innovations, actual gales of technological change may emerge. The wider the access to local pools of knowledge, the greater the likelihood that firms are induced to react. The larger the number of firms that react and the better the access conditions to external knowledge, the stronger the chances that their reactions are creative: technological change becomes a generalized and collective process (Arthur, 1989, 2009).

In such a context innovation is an emergent property that takes place when complexity is 'organized', that is, when a number of complementary conditions enable the creative reaction of agents and make it possible to introduce innovations that actually increase their efficiency. The dynamics of complex systems is based upon a combination of agents' reactivity, caught in out-of-equilibrium conditions, and features of the system in which each agent is embedded in terms of externalities, interactions and positive feedback that enables the generation of localized technological change and leads to endogenous structural change (Anderson et al., 1988; Arthur et al., 1997; Lane et al., 2009).

Innovation is the endogenous result of system dynamics: it does not fall from heaven, as standard economics suggests; neither is it the result of random variation, as some evolutionary approaches – with their strong Darwinistic traits where mutation takes place randomly – consistently contend. Agents react and succeed in their creative reactions when a number of contingent external conditions apply at the system level. Innovation is the result of the collective economic action of agents:

> innovation is a path-dependent, collective process that takes place in a localized context, if, when and where a sufficient number of creative reactions are made in a coherent, complementary and consistent way. As such innovation is one of the key emergent properties of an economic system viewed as a dynamic complex system. (Antonelli, 2008: 1)

Appreciation of the systemic conditions that shape and make innovations possible, together with their individual causes leads to the identification of innovation as an emergent property of a system. Our approach provides a solution to the conundrum of an intentional economic action whose rewards are larger than its costs. This can happen only if the complexity of the system is appreciated. The introduction of innovations that make it possible to enhance the productivity and efficiency of the system can in fact take place only as the emergent property of an organized system complexity, and in turn organized complexity, is explained as an endogenous and dynamic process engendered by the interactions of rent-seeking agents trying to cope with the ever-changing conditions of their product and factor markets (Antonelli, 2009, 2011).

Architectural and Institutional Trade-Offs

In this context, because of the twin character of knowledge as output of research and input in further knowledge, two knowledge dissemination trade-offs take place. The first relates to the structure of intellectual property rights (IPR) regimes; the second relates to the distribution of knowledge generation activities in economic, regional and knowledge space. Let us analyse them in turn.

The intellectual property rights trade-off
The structure of IPR regimes – the scope of patents, their duration, assignment procedures and their exclusivity – plays a crucial role. Strong IPR regimes increase the appropriability of technological knowledge as they limit leakage of information and delay uncontrolled knowledge

dissipation. Innovators can secure for a longer period of time the benefits stemming from the generation of new technological knowledge and the introduction of new technologies. Strong IPR regimes increase the innovators' opportunities to exploit technological knowledge. Consequently strong regimes enhance incentives to the generation of new knowledge and hence help increase the amount of resources that would be committed to it. Strong IPR regimes, however, reduce both the static and the dynamic efficiency of economic and innovation systems. Strong regimes increase the duration of monopolistic power in the product markets and the appropriation of consumer surplus by innovative suppliers. These regimes, however, reduce the dynamic efficiency of innovation systems because they prevent and delay access to existing knowledge as an input in the generation of new knowledge, and hence reduce the efficiency of the recombination process that leads to the generation of new technological knowledge. The combined effect of strong property rights regimes in fact is to increase the incentives to generate research, and hence resources; but there is a reduction in their efficiency because at each point in time available knowledge cannot be used to recombine and generate new knowledge and must be reinvented. Strong IPR regimes risk increasing the replication of research efforts and slowing the pace of recombinant generation of technological knowledge. This knowledge trade-off requires the fine-tuning of intellectual property rights with the identification of the proper mix of protecting appropriability on the one hand and disseminating available knowledge on the other.

The architectural trade-off
The architectural characteristics of the network of interactions that qualify each economic system have powerful consequences for the actual capability of each economic agent to generate new technological knowledge. The distribution in regional and knowledge space of knowledge generation activities has important effects. Because of the pervasive role of external knowledge in the recombinant generation of new technological knowledge the regional concentration of knowledge generating activities may increase the pace of technological advance. Proximity, in fact, helps the identification of useful external knowledge, and hence reduces search and exploration costs. Proximity in regional space helps reduce the risks of opportunistic behaviours because of increased interaction, and hence helps limit transaction costs; and, finally, proximity increases the homogeneity of codes and favours the absorption of external knowledge. Excess concentration may favour the forging ahead of small but effective clusters of highly innovative groups of firms strongly interconnected and able to interact quickly. At the same time, however, excess concentration might be

identified where the rest of the system is cut off from the flows of creative interactions and the dissemination of new knowledge is delayed. Excess concentration risks reducing knowledge variety and the related opportunities for knowledge recombination. The dissemination of knowledge generating activities may help stimulate the recombinant generation of new knowledge because of the wider participation of a larger number of heterogeneous agents in the collective endeavour that leads to the generation of new knowledge. Once more it is clear that a knowledge trade-off between the concentration and dissemination of knowledge generating activities takes place with important policy implications for the best allocation of additional research resources and activities through regional space (Page, 2011).

Agent-based models can help structure the dynamic properties of the system in a rigorous framework of analysis so as to provide a context in which the implementation of simulation techniques can exhibit the different results of alternative structures of knowledge interaction mechanisms and IPR regimes.[2] This exercise can contribute to an approach that adapts complex system dynamics to economics where technological change is the central engine of the evolving system dynamics and the result of the creative response of intentional agents embedded in an evolving architecture of market, social and knowledge interactions (Aghion et al., 2009; Terna, 2009).

Modelling an economic system where technological change can take place implements the basic intuitions of complexity theory and innovation economics. The model will enable firms to identify proper solutions to the two knowledge trade-offs identified with respect to the structure of IPR regimes and the regional distribution of knowledge generation activities.

Let us now turn our attention to analysing the building blocks of our agent-based simulation model. The following section shows how the use of the basic tools of ABM can implement a rigorous representation of the dynamics of a fully fledged economic system where agents can generate technological knowledge and technological innovations, provided a conducive architecture of network interactions and an effective IPR regime are implemented.

THE SIMULATION MODEL

The system of interactions and transactions that qualify the simple but articulated economic system outlined in the previous section can be explored by means of ABM in order to investigate the dynamics of the innovation process at the system level. ABM provides us with the opportunity to explore the full range of implications of a multilevel structure of interactions and transactions as framed in the previous section and to take into account the outcomes of the decisions taken by each heterogeneous agent (Pyka and Werker, 2009; Terna, 2009).

The ABM implemented in this section operationalizes, through interactions among a large number of objects representing the agents of our system, a typical complex process characterized by the key role of Marshallian externalities and augmented by the Schumpeterian assumption that firms can try to innovate according to their performance levels and local contexts (Dawid, 2006).

The model assumes that firms are boundedly rational but endowed with procedural rationality enriched by the ability to react and innovate when and if a number of external circumstances are provided. The rationality of their behaviour is objective as opposed to subjective. Firms do in fact react to the dynamics of both product and factor markets, but never maximize. Their reaction includes the possibility to innovate instead of merely adjusting quantities to prices.

In the ABM supply and demand meet in the market place; production is decided *ex ante*; firms try to sell their products in the market, where customers spend their income. The matching of demand and supply sets temporary prices that define firms' performance. According to the levels of their performance and the availability of external knowledge firms can fund dedicated research activities to try to innovate (Lane et al., 2009).

In the simulation, heterogeneous firms produce homogeneous products sold into a single market. In the product markets households expend revenues stemming from wages (including research fees) and the net profits of shareholders. In the input markets the derived demand of the firms meets the supply of labour provided by workers, including researchers. For the sake of simplicity, no financial institutions have been activated, and nor can payments be postponed. Shareholders supply the whole capital of the firms and all the commercial transactions are immediately cleared (Figure 2.1).

Market clearing mechanisms based exclusively upon prices maintain perfect equilibrium between demand and supply. Such equilibrium is ensured for both the product and the factor market: the quantities determine the correct price to ensure the whole production is sold. No friction or waiting times are simulated; factors are assumed to be immediately available. Here joint reference to the Marshallian and Schumpeterian legacies plays a key role in understanding the working of such markets. At each point in time the market equilibrium is typically Marshallian as opposed to Walrasian. Here exchanges occur after production. Production has been taking place according to the plans based upon the expectations, beliefs

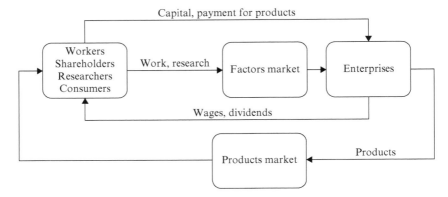

Figure 2.1   Fluxes in the simulated economy

and technological competence of each agent. For each transient market price, because agents are heterogeneous, some make profits and others incur losses. Following the Schumpeterian traits of our model, however, no convergence can take place as long as firms introduce innovations and hence keep changing the attractors.

The production function is very simple in order to avoid matters related to different production processes, input availability, warehouse cycles and so on: output depends exclusively on the amount of employed work and productivity. Both labour and productivity vary among firms: labour depends on the entrepreneur's decision about the growth of the production; productivity is a function of the technological level the firm achieved through innovation.

The entire output is sold in the single product market, where revenue equals the sum of wages, dividends and research expenses and price depends on liquidity. According to temporary price levels, profits are computed as the difference between income and expenses, no taxes are paid and no part of the profit can be retained in the enterprise. Shareholders will either receive the profits or reintegrate the losses.

Heterogeneous firms are localized in an economic structure represented as a regional and technological space. Both spaces are managed as grids divided into cells, each of which can host an unlimited number of firms. Position in the regional grid determines the neighbourhood in which firms can observe their competitors, comparing results. The position of each firm in the technological grid measures its productivity and defines the possibility to access quasi-public knowledge. The distribution in the two space dimensions is not consistent: firms technologically very close could be positioned in far distant cells of the regional space, and vice versa. In this way the absorption of technological knowledge spilling from firms based in regional and technological proximity may enable the introduction of an innovation with positive effects in terms of productivity growth.

The localization of agents in both space dimensions is the result of their past activities, and yet it can be changed at each point in time. The results obtained during a production and consumption cycle influence the strategies the agents will take during the next cycle. Hence the dynamics of the model is typically characterized by path dependence: the dynamics in fact is non-ergodic because history matters and irreversibility limits and qualifies the alternative options at each point in time. At each point in time, however, the effects of the initial conditions may be balanced by occasional events that may alter the 'path', that is, the direction and pace of the dynamics (David, 2007).

The firms in the model, in fact, are always comparing their perform-ance in terms of profits to their neighbours' average results, with the difference between own figures and neighbours' averages increasing the motivation to innovate. Transparency is clearly local: the radius within which firms can observe the conduct of other firms is limited accord-ingly by a parameter value. Beyond that radius information is scarce and costly. The further profitability is from the average, the deeper the out-of-equilibrium conditions. Firms can innovate if the results are below average, to improve their performance, as well as when results are above average, to take advantage of abundant liquidity and reduced opportunity costs for risky undertakings. Innovation is viewed as the possible result of intentional decision-making that takes place in out-of-equilibrium conditions. The further away the firm is from equilibrium, the stronger the likelihood for innovation to take place. Hence we assume a U-relationship between levels of profitability and innovative activity, as measured by the rates of increase of total factor productivity (Antonelli and Scellato, 2011).

Summarizing, a firm increases its motivation to innovate each time its performance is found to be far enough from the average. Such a motivation becomes stronger and stronger if the enterprise's relative position remains outside a band for several and consecutive production cycles: after a parametrically set number of consecutive cycles the enterprise performs an innovation trial.

Simulation of the Innovation Process

ABM enables us to explore in detail the innovation process and the role within it of the external factors that shape the recombinant generation of technological knowledge. At each point in time firms can react so as to try to increase their productivity. Hence they can move and change their regional and technological localization by means of research costs. The research costs are directly related to the actions performed by each firm to innovate, namely:

- mobilizing internal slack competence;
- absorbing external knowledge spilling from neighbours;
- moving to other locations in order to exploit more developed neighbours.

We assume a sequence of innovative steps. At first firms try to mobilize their own internal slack competence. Firms with insufficient potential try to absorb the external technological knowledge spilling from a neighbour and, if knowledge absorption is not possible, they can move randomly to another location in the physical space. Let us consider these actions in turn.

Mobilization

Firms can mobilize their internal slack competence accumulated by means of learning processes. The firms in the model are endowed with the ability to learn better ways of performing their production cycles. After every production cycle firms acquire and cumulate some technological potential. Such potential can be transformed into actual innovation only by means of appropriate research activities and access to external knowledge. Firms are able to build up competence by means of learning processes. The accumulation of experience proceeds at a specific internal 'learning rate' that is biased by the impact of external 'learning factors' that reflect the competence level of the enterprises' locality, measured as the average productivity of neighbouring enterprises. The competence can be transformed into real innovation when a parametrical threshold is reached, at a cost. Because the internal slack competence is seldom sufficient to support the recombinant generation of new technological knowledge, and hence the actual introduction of a productivity-enhancing innovation, firms explore their local technological and regional space and try to access and absorb the knowledge of their neighbours (March, 1991).

Absorption

Absorption enables firms to take advantage of technology introduced by other firms; because of absorbing costs, however, this is not free. Effective access to external technological knowledge requires substantial resources in exploration, identification, decodification and integration into the internal knowledge base (Cohen and Levinthal, 1989, 1990). Moreover, because of bounded rationality, firms can observe only other firms in a certain neighbourhood whose extension depends on a 'view' parameter; this value limits the number of positions all around the agent it can explore. Due to the fact that the simulated world is managed as a grid the agent's position can limit this view: agents in a corner have less possibility to observe than others located in the middle of the grid, whereas agents in a very crowded neighbourhood have more information than isolated firms. Note that a single position on the grid could contain several agents, so simply by exploring its cell an agent may find other firms to observe.

The view parameter determines only the number of cells the agent can access; the real number of other firms it can observe depends on the evolution of the agent's distribution and constitutes an emerging phenomenon that continuously evolves during the simulation. When the agent is located near the end of the grid its capability falls dramatically.

A major constraint to the possibility to take advantage of and absorb others' technologies is represented by intellectual property rights. In order to model a credible IPR regime we allow enterprises to patent their technology and hence retain exclusive exploitation rights for a certain number of cycles (Reichman, 2000). By observing other firms each firm knows the latest technological level they apply that is not covered by a patent licence. The key parameter 'patent expiration' (pe) is used to test different scenarios; its value determines the number of production cycles each innovation remains hidden from competitors. It is plausible to expect that the longer the patent period value of the patent expiration parameter, the higher will be the research effort: unless enterprises are given the exclusive possibility to exploit the research results, no private firms would be interested in investing money because their discovery would be immediately available to competitors. In the model, even with patent expiration equal to zero, the new technology is exploited exclusively by the innovating enterprise for almost one cycle.

Observed technologies can be absorbed only if the distance between them and the firm's own is less than a parametrical value, the so-called 'knowledge absorption threshold'. This limitation has been introduced to avoid dramatic jumps in the productivity of firms that would not be plausible. Knowledge absorption has a cost equal to the named distance. Because the possibility to observe neighbours depends on the position of each enterprise in the physical space, when knowledge absorption gives poor or null results enterprises could decide to move to another location in order to meet better technological conditions.

Relocalization
The third way to improving productivity levels consists in moving around the physical space in order to reach more interesting neighbours. When

the mobilization of competence and knowledge absorption are not viable solutions, firms can try to move randomly to another location in the hope of finding better-developed zones. Movement is limited by a parameter called 'jump', whose value determines the maximum number of cells the firm can go through (vertically, horizontally, backwards or forwards). The effective number of cells the enterprise will move is determined randomly in this range, which constitutes a von Neumann neighbourhood. Moving costs are directly related to the distance between the original and the new location.

Here we see how the structure of the system influences in several ways the innovation chances of an enterprise: learning is faster for firms that operate in a well-developed neighbourhood, and imitators have better opportunities to observe and copy if they operate in a crowded and technologically advanced environment (Ozman, 2009).

Firms are endowed, at the start of the simulation, with a competence and a technological level randomly chosen for each in the lowest quarter of possible values, following a uniform probability distribution. The simulations started with low-skilled firms with uniform distribution among them, both to give each firm the possibility to express its own development path and a similar starting situation to analyse the different development paths.

In the real world, knowledge centres such as universities, technical and management schools and so on are located unevenly in the geographical territory, with clear effects. A large amount of evidence confirms that firms operating in geographical regions with a high density of such organizations have better chances to access higher levels of knowledge. To introduce these aspects in the simulation model we have represented geographical regions by physical spaces where competence is distributed following different configurations, from full concentration in a limited space to well-disseminated distribution. Knowledge centres are represented by firms with a very high technological level (so-called 'genius') whose initial knowledge endowment is randomly tossed within the highest quarter of the possible values, whereas 'normal' agents are given values in the lowest one.

Neighbours can take advantage of the external knowledge spilling from the 'genius' firm within the boundaries of the knowledge absorption threshold value set up for the simulation. Hence the higher the knowledge absorption possibility, the stronger is the influence of the genius on its neighbours. Patent duration does not slow the effect because the initial knowledge is assumed to be an old and public one.

In order to experience different scenarios the number of genius firms is parametrically managed and could be set to zero to exclude this effect. The distribution in space of agents is random at the beginning of the process, but it becomes fully endogenous as agents are credited with the ability to move in regional space, searching for access to external knowledge spilling in the proximity of 'genius'. Hence the dynamics of the regional distribution of agents exhibits the typical traits of path dependence. The process is non-ergodic but not past dependent: small variations can exert important effects in terms of the emergence of strong clusters or, conversely, progressive dissemination in space (D'Ignazio and Giovannetti, 2006; Antonelli, 2008).

RESULTS OF THE SIMULATIONS

The strength of the ABM consists in the possibility to assess in a coherent and structured framework the systemic consequences of alternative structural configurations of the system properties. Simulation techniques allow us to explore the outcomes of different hypotheses concerning key issues of the model within a structured and consistent framework that takes into account the full set of direct and indirect effects of the interactions of agents (Pyka and Werker, 2009).

The results of the simulation confirm that the model is consistent and able to mimic the working of a complex system where rent-seeking agents react to the changing conditions of the product and factor markets. Hence the results confirm that the set of equations is able to portray the working of a complex system based upon a large number of heterogeneous agents on both the demand and the supply side that are price takers in product markets. Markets clear with temporary equilibrium price. Replication of the temporary equilibrium price in the long term confirms that the model is appropriate to explore the general features of the system when the reaction of firms is adaptive and consists in price to quantity adjustments. In the extreme case where firms cannot innovate for the lack of internal competence to be mobilized and external knowledge to be absorbed, the system mimics effectively the working of static general equilibrium in conditions of allocative and productive efficiency, but with no dynamic efficiency. The markets sort out the worst-performing firms and drive prices to the minimum production costs. This result is important because it confirms that static general equilibrium is the simple and elementary form of complexity that takes place when agents cannot innovate. As soon as agents try and succeed in their reaction to changing market conditions with the introduction of innovations, the equilibrium conditions become dynamic and all the key features of the system – such as prices, quantities, efficiency and structure – keep changing (Antonelli, 2011).

Innovation is effectively an emerging property of the system because it takes place when the external conditions and the structure of the system provide access to the external knowledge that is crucial to feed the effective recombinant generation of new technological knowledge, and hence the actual introduction of productivity enhancing innovations by firms that try to cope with the changing conditions of the system by doing more than merely adjusting prices to quantities.

Access to external knowledge is necessary to achieve the effective recombinant generation of new technological knowledge and to eventually introduce new technologies. The structural characteristics of the system in which firms are embedded are crucial to enable the reaction to become creative, and hence to introduce innovations that increase their productivity. The simulations provide key information about the two knowledge trade-offs and enable us to assess the systemic effects in terms of dynamic efficiency of alternative configurations of the IPR regimes and architectures of the network interactions. We have explored the consequences of two sets of hypotheses: i) the effects of different durations patents; and ii) the effects of different architectural properties of the system in terms of the distribution of firms with high levels of technological competence.

The effective recombinant generation of technological knowledge and the consequent introduction of technological innovations is tracked and quantified in terms of productivity growth, measured as the ratio of input to output. Firms that are able to take advantage of knowledge externalities to successfully generate new technological change, and hence to introduce better technologies, will experience an increase in the general levels of efficiency of their production process and will experience higher mark-ups with evident positive consequences on productivity levels.

The changes in productivity levels affect the system dynamics not only in terms of average growth rates but also in terms of variance. Growth-cum-technological change is far from a steady increase. On the contrary, it exhibits fluctuations that are typical of the long-term Schumpeterian process of creative destruction. Occasionally the majority of firms incur major losses due to mismatch between their current cost conditions and the performance of a few radical innovators able to introduce breakthrough innovations. In a typical Schumpeterian process we see that the introduction of radical innovations engenders occasional phases of decline in output. It is interesting to note that the fluctuations are sharper when the pace of technological change is higher and, more specifically, in the configurations of spatial distribution and appropriability regimes that increase the rates of introduction of technological innovation, and hence of productivity growth.

Let us now consider in turn the alternative results that are obtained with different structural configurations of both IPR regimes and the spatial distribution of firms.

The First Knowledge Trade-Off: Intellectual Property Rights Regimes

The first question the simulation has been employed to investigate concerns the role of patent protection in promoting and sustaining innovation. The well-known IPR trade-off can now be investigated (Harison, 2008; Vandekerckhove and De Bondt, 2008).

Intellectual property rights enable firms to secure exclusive rights on the technological knowledge they have generated. By means of IPR enterprises can exclude competitors from the exploitation of such new technologies and consolidate an effective competitive advantage. At the micro-level patent protection reinforces the motivation to innovate, giving an enterprise the possibility to exploit its own innovation in an exclusive way (hereafter 'reinforcing effect').

Moving from our basic assumption that the introduction of innovations builds upon the recombination of existing knowledge, it is clear that patent protection has a negative effect: the longer the protection lasts, the slower the new technologies can spread among firms (hereafter 'slowing effect') (Gay et al., 2008). This research investigates both effects, focusing on the influence they have on the innovation process. The simulations were run using the following set-up:

- All the firms (agents) operated in a common market and district.
- All the firms started from a similar technological level, randomly tossed into the first quarter of the achievable technologies following a uniform distribution.
- Each firm was given high capability to observe neighbours and absorb external knowledge.
- The unique parameter that varied among the simulations was 'patent expiration' – the time, in production cycles, that a new technology was owned by the innovator and not available to the other agents in the system.
- The probability of firms trying to innovate even if their results are similar to their neighbours' is positively correlated to patent expiration if it is less than 100; for values greater than 100 no innovation at all is pursued by the firms unless their results differ greatly from their neighbours'.

Two sets of experiments were executed, both based upon observation of the average productivity level the agents achieved after a determined number of production cycles. In the model productivity is positively correlated to the technology, so the more a firm innovates the higher is its productivity. By observing the dynamics of the productivity it is possible to study the effects of the institutional and regional context on the innovation strategy of the firms.

The first set of experiments consisted in benchmarking the innovation to explore the difference in the results in terms of productivity levels obtained with several different patent protection durations and a benchmark figure represented by the productivity level the agents achieved with patent expiration set to one. To ensure the results were robust and systematic, each simulation was run ten times by varying, for each run, the seed employed to generate pseudo-random numbers; the result of each experiment was computed as the average of the ten runs. The second set of experiments consisted in correlating innovation and patent expiration: 50 simulations were run, varying each time both the random seed and the patent expiration parameter; the value was randomly tossed following a uniform distribution into the interval:]1,255[. The described approach ensured both the robustness of the results and the independence of the parameters set up from any researcher's mental schemata.

Benchmarking the productivity

Figure 2.2 shows the average results obtained in five different experiments based upon diverse values for the patent expiration parameters. The results of the first one (patent expiration set to one) constitute our benchmark. Each experiment consisted in running for several times (ten in this case)

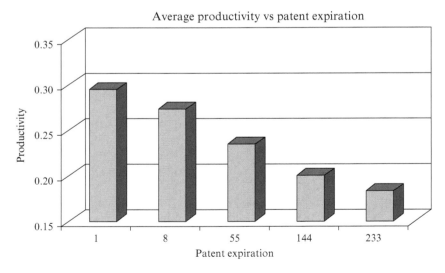

Figure 2.2   Histogram representing the results of the simulations

a simulation covering 500 whole production–consumption cycles, with a determined value for the patent expiration parameter and different random seed. The results of each simulation have been summarized by means of the average productivity value computed tacking the reached values of each firm into the population of the model.

The distribution of those average values exhibited a very low variance, allowing their use as the representative values and suggesting that the results were robust and fully independent of the different random number distributions generated for each simulation starting from a diverse seed.

Figure 2.2 shows that the four scenarios 8, 55, 144, 233 were not able to achieve the benchmark (scenario 1); because the productivity level directly depended on the achieved technological level, it would mean that the reinforcing effect was systematically weaker than the slowing one. In sum, the results confirm that the stronger the IPR protection (the more extended in time the patent protection was), the slower the innovation process proceeded.

Figure 2.3 shows in more detail that in all the simulations the results were systematically higher the lower the patent expiration.

Figure 2.4 better shows the trend of the phenomena by drawing on the minimum and maximum results obtained in each experiment.

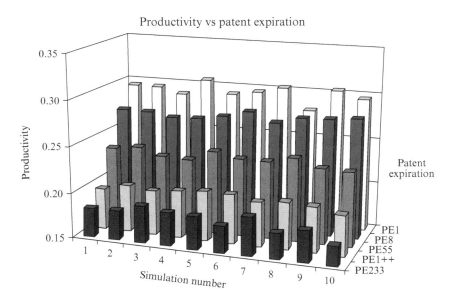

Figure 2.3   Results reached in each of the ten simulations

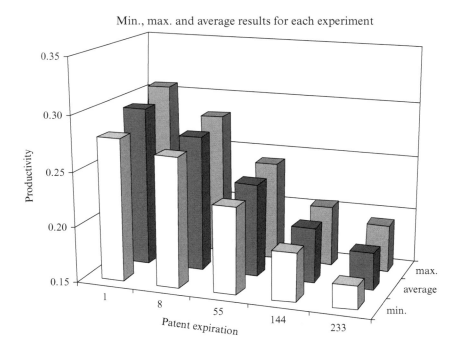

Min., max. and average results for each experiment

Figure 2.4   Results of the experiments related to the patent expiration parameter

Correlating innovation and patent expiration

Figure 2.5 shows an early correlation between patent duration and productivity levels that the simulated economy reaches by grouping each set of five simulations, picking the first, second and so on of each experiment under different values of patent expiration. The results obtained by running 50 simulations, 500 production cycles long with random values for patent expiration demonstrate the existence of a negative correlation between patent rights and innovation.

The longer the patent right, the less the productivity level grows, as graphically illustrated by Figure 2.6. The obtained correlation index is about −0.9; the distribution of the obtained results shows a remarkable relative difference between the best case (patent expiration = 6) and the worst (patent expiration = 214).

Figure 2.6 and Figure 2.7 illustrate, respectively, the distribution of average productivity values and distribution of the relative difference between each value and the worst case. The productivity difference (dp) was computed as $dp_i = p_i/min(p) - 1$, where p_i represents the

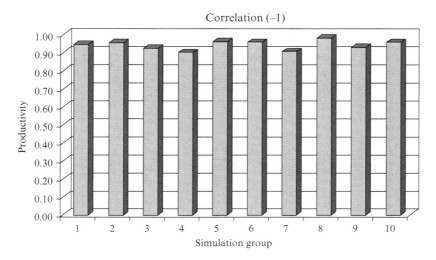

Figure 2.5   Correlation between patent expiration and productivity

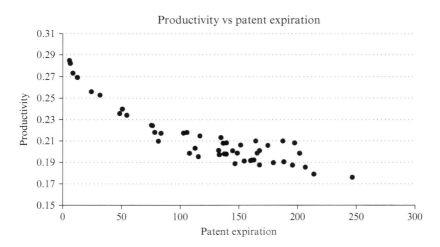

Figure 2.6   Distribution of average productivity with different patent expiration parameters

productivity of the i-th experiments and min(p) the minimum productivity level achieved in all the experiments. A similar algorithm was employed to compute the patent expiration difference (dpe): $dpe_i = pe_i/max(pe) - 1$.

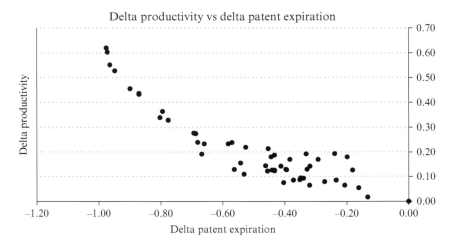

Figure 2.7   Distribution of relative differences versus the worst case

The Second Knowledge Trade-Off: The Regional Dissemination of Knowledge

The second issue addressed by the simulations concerns the role of the distribution in regional space of knowledge generating institutions such as research laboratories, universities and so on in promoting and sustaining innovation.

We want to test the hypothesis that the dissemination of knowledge favours system growth. This very first stage of the research has been focused on the influence that different architectural distributions of the knowledge producers (KPs) have on the dynamic of the innovation process. The distribution in regional space of KPs is a valuable source of the recombinant generation of new technological knowledge as they provide the opportunity to all the other co-localized agents to access part of their proprietary knowledge in the form of knowledge spillovers (Ozman, 2009).

In order to maintain the model at a useful level of simplicity, the KPs have been dummied by some highly evolved firms whose distribution will affect the possibility of other firms taking advantage of the technological knowledge spilling from them.

The distribution of knowledge has been simulated by inserting a small number of firms endowed with a high level of technological knowledge (genius) into an environment populated by a wide set of less-developed firms. The different distributions and numbers of genius firms have been studied in several scenarios, that is, under diverse set-up of some basic parameters that determine the quality of information available, the limits of the physical relocation, the ability to observe and copy others' strategy and so on.

Four different distributions for knowledge producers have been compared by observing their effects on the evolution of productivity; to ensure the distributions were stable knowledge producers were not allowed to change their position in the physical space. In the four spaces we find 250 normal firms and a certain number of knowledge-intensive ones (KPs). In each space the distribution of the high-tech firms is set up as follows:

- one high-knowledge district (hkd): all the KPs are placed very close together in a small area at the centre of the space;
- two hkds: the total number of KPs is split between two areas, one at the centre of the right upper quarter of the space and the other at the centre of the left lower one;
- four hkds: the KPs are distributed around four points at the centre of each quarter the whole space could be divided into;
- no hkds: each KP is assigned a random position in the space and lives alone.

The basic population of each region (about 250 agents due to the fact each agent is assigned a random space tossed following a uniform distribution) is randomly spread in the space.

Each set of experiments has been based on a different combination of four parameters (scenarios), designated as follows:

- Optimum: the scenario devoted to recreating the theoretical condition of perfect information and mobility. Here agents have a large view, knowledge is fully available and moving is always possible.
- Typical: here the capabilities of the firms are limited to plausible amounts in order to take into account the typical limits existing in the real world.
- Mixed: the parameters are randomly set up for each simulation, choosing their values in an assigned range that includes 'typical' values.

For each scenario three different experiments were done using 4, 16 and 64 KPs for each space. By varying the number of KPs the difference between each KP distribution model could be differently stressed: with four KPs for each space there is little difference between their diverse distribution and, practically, four hkds is one of the possible distributions of the no

Table 2.1   Parameter configurations for each experiment

Scenario	Parameters				Number of KPs	Experiment
	View	Jump	Imitation threshold	Patent expiration		
Optimum	15	50	999	1	4	Optimum 4
					16	Optimum 16
					64	Optimum 64
Typical	4	4	4	5	4	Typical 4
					16	Typical 16
					64	Typical 64
Mixed	]0,8[	]0,8[	]1,9[	]1,15[	4	Mixed 4
					16	Mixed 16
					64	Mixed 64

hkds scenario. The more the number of KPs is increased, the greater the difference between the four distributions.

Each experiment was repeated 50 times, always changing the random number distribution to simulate different dynamics and validate the robustness of the obtained results. Random numbers were used to simulate some decisions, to pick up neighbours spilling relevant external knowledge and to determine in which direction and how far to move. For the mixed scenario random numbers were used to toss the parameters' value, each within the appropriate range, as illustrated in Table 2.1 where parameters for each scenario and simulation are shown.

At the end of each experiment the average productivity level for each region and for the whole population have been computed; at this very first stage of the research these were the only data it was decided to concentrate on.

Since the initial endowment of the firms in each region was set to the same amount, the market was unique both for factors and products. It is possible to assume differences in the reached level of productivity were mainly due to the different distribution of the KPs, as shown in Figure 2.8.

Results of the Optimum Scenario

The optimum scenario was set up to validate the model under the classic assumption of perfect information and mobility: provided that each regional space is simulated by a square lattice 100 cells wide, jumping in each direction of 50 cells means having perfect mobility. In addition, because the maximum distance between the worst and the best technology

Different distributions of the KPs

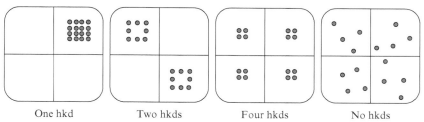

| One hkd | Two hkds | Four hkds | No hkds |

Figure 2.8   Configuration of the spaces for the simulations

Table 2.2   Synthesis of results obtained by running the optimum scenario

Experiment	Function	One hkd	Two hkds	Four hkds	No hkds
Optimum 4	Average	0.959782	0.939995	0.875459	0.858535
	Variance	0.001633	0.013787	0.031749	0.026233
Optimum 16	Average	0.991779	0.993152	0.984252	0.971355
	Variance	0.000074	0.00002	0.000630	0.001899
Optimum 64	Average	0.994997	0.994967	0.994451	0.993284
	Variance	0.000000	0.000000	0.000000	0.000022

has been limited in these simulations to 200 and knowledge absorption threshold of 999 means that each technology could be copied. Patent expiration set to one means that each adopted technology becomes quasi-public in the successive production cycle, so each technology could be copied as soon as it has been adopted.

The value of the view parameter would have been set to 50 too, as for the jump one, but 15 was demonstrated to be enough to allow a good circulation of information and guarantee the majority of the enterprises reached the higher technological level in a very short time.

Under the optimum conditions, the concentrated distributions of KPs, such as the one hkd and two hkds seem to give some advantages, as shown by the results briefly summarized in Table 2.2. Here are reported, for each experiment, the average results (first row) obtained during 50 runs, with different random distributions, each of them 250 whole production cycles long; the variance is reported too, in the second row.

With high levels of information quality, mobility and capability of firms to absorb technological knowledge from each other, and no patent protection, the concentrated distribution of knowledge centres seems to give

Experiment optimum 64

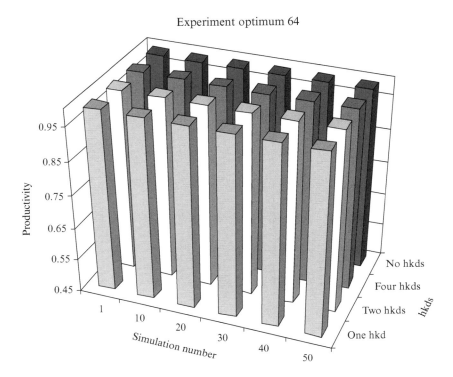

Figure 2.9   Results of several simulations of the experiment optimum 64

better results than the disseminated one, even if the advantage becomes smaller and smaller when the number of KPs grows.

Starting from the scenario with only four KPs the disseminated region reached only 0.85 productivity after 250 production cycles, whereas the fully concentrated one reached 0.95, with an advantage of about 0.1; but this difference fell to 0.02 and 0.001 respectively with 16 and 64 KPs.

The trend shown by the average values systematically appears in each single simulation, as Figure 2.9 illustrates for optimum 64: the graph reports the final results of each of the 50 simulations.

Results of the Typical Scenario

This configuration set was obtained by giving the four parameters realistic and plausible values. The regional neighbourhood of each firm was presumed to be 64 cells wide, about 1/100 the whole extension of the simulated world, where each cell was able to host more than one enterprise. Assuming this neighbourhood to be the maximum extension a firm would have been able to reach, the possibility to move was limited to the same amount.

Innovation cannot be done too fast; the absorption and recombination of external technological knowledge implies the modification of products and production processes and upgrading of the skills of the staff; it is not plausible that an enterprise can absorb unlimited amounts of external technological knowledge. The limit of four represents 1/50 of the maximum technology a firm can reach in the whole evolution, and 400 times the ability each enterprise is assumed to acquire each cycle by means of 'learning by doing'.

It is also plausible that new techniques could be protected by licences. Usually technical patents last for five years because each step of the simulation is assumed to last one year, so the expiration of patent rights has been set to five. Practically each firm can observe and absorb other technologies only if they are five cycles old. All these limitations reduced the speed of evolution, so experiments for this scenario were based on simulations 1000 cycles long, even though the enterprises reached productivity levels less than obtained in the (non-realistic) optimum scenario. The interesting result is that, under more realistic conditions relevant indications about the better distribution of KPs seem to appear, as in Table 2.3, where are shown the average results of 50 runs for each experiment using the typical scenario.

In all the three set-ups of KPs, the disseminated distribution provides better results and the distance becomes higher the higher the number of KPs.

Analysing the four regions it is evident that the more the KPs are spread, the better the results become: the advantage grows significantly, passing from the one hkd scenario region to the no hkds one, reaching, for 64 KPs, 0,16. Figure 2.10 shows the results obtained during the 50 simulations for the experiment typical 64.

More disseminated distribution of the KPs seems to be more effective in facilitating innovation and in promoting technical progress. A plausible explanation could be that more disseminated distribution allows a major

Table 2.3   Synthesis of results obtained by running the typical scenario

Experiment	Function	One hkd	Two hkds	Four hkds	No hkds
Typical 4	Average	0.455182	0.480327	0.501641	0.508746
	Variance	0.005142	0.007908	0.008847	0.007251
Typical 16	Average	0.656397	0.695474	0.738526	0.78994
	Variance	0.003172	0.001373	0.001982	0.001805
Typical 64	Average	0.796893	0.844447	0.906629	0.957159
	Variance	0.001475	0.000630	0.000102	0.000059

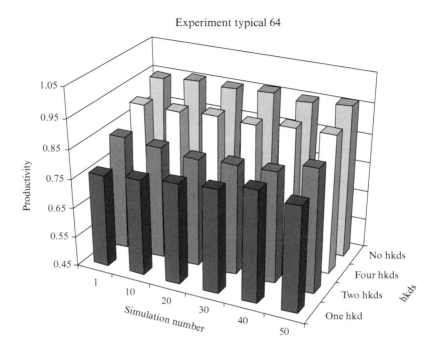

Figure 2.10   Results of several simulations of the experiment typical 64

number of firms to access knowledge. Similar configurations, like four hkds and no hkds in the presence of four KPs only, gave very close results, confirming this explanation.

Results of the Mixed Scenario

The mixed scenario was built to test the results obtained into the typical one. Here the parameters set-up is always changing; values are randomly tossed in ranges that are distributed around the typical parameters value.

The results, reported in Table 2.4, confirm those obtained by running the typical scenario, so the previous reasoning about the importance of a disseminated distribution for KPs seems to be reinforced, as well as the observation about the similarity between the distribution four hkds and no hkds in the presence of four KPs only.

The difference among the four distributions is less strong, due to the fact the combination of parameters allowed configurations closer to the optimum scenario than the typical one. The phenomenon is clearly shown in Figure 2.11.

Table 2.4   Synthesis of the results obtained by running the mixed scenario

Experiment	Function	One hkd	Two hkds	Four hkds	No hkds
Mixed 4	Average	0.444593	0.456321	0.476576	0.474556
	Variance	0.023025	0.025162	0.026424	0.02444
Mixed 16	Average	0.573829	0.599636	4.507638889	0.697282
	Variance	0.023182	0.01804	0.019603	0.018044
Mixed 64	Average	0.771741	0.799501	0.84245	0.895232
	Variance	0.023435	0.019453	0.017576	0.014354

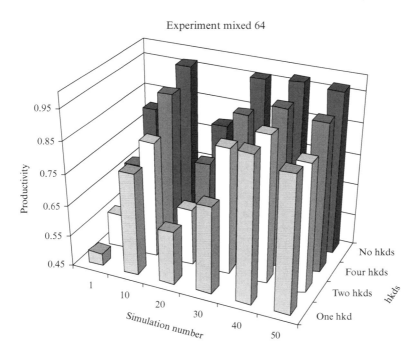

Figure 2.11   Results of several simulations of the experiment mixed 64

Taking advantage of the array of experimental configurations that agent-based simulations offer, we have generated a wide set of alternative scenarios. For a comparative summarization it is possible to refer to Figures 2.12 to 2.14, where the data shown above are mixed in bar diagrams.

Whereas in the optimum scenario results are very similar for the three different distributions, the advantage of the no hkds distribution is clear in the typical and mixed scenarios.

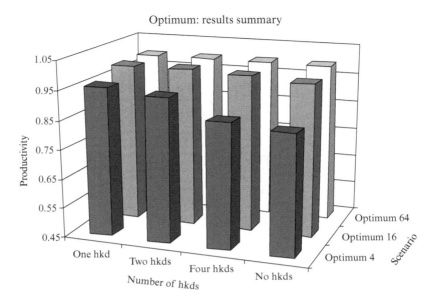

Figure 2.12   Comparison of results of the optimum scenario

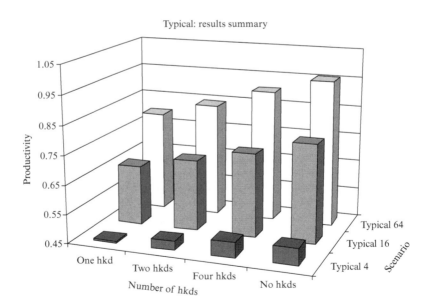

Figure 2.13   Comparison of results of the typical scenario

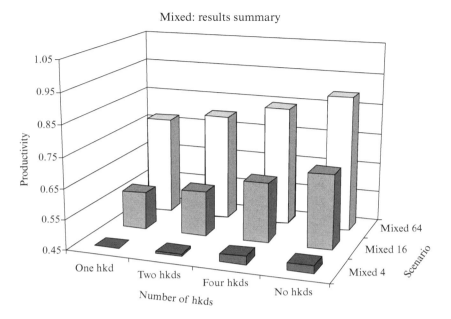

Figure 2.14   Comparison of results of the mixed scenario

The bar diagrams show also that the performance of the disseminated distributions are better the higher is the number of KPs, reinforcing the argument about the similarity of distributions in the presence of few KPs.

POLICY IMPLICATIONS

The implications for research and innovation policy are important: better access conditions to technological knowledge and better dissemination of existing technological knowledge enable firms to better find their way toward technological enhancement so as to become more competitive and profitable. Let us consider them in turn.

Intellectual property rights regimes should be designed so as to increase the possibility for imitators and users of external knowledge to take advantage of existing proprietary knowledge. The implementation of non-exclusive IPR might favour the dissemination of technological knowledge. The enforcement of compulsory royalty payments for all use of proprietary knowledge should prevent the reduction of appropriability conditions, and hence the decline of incentives to funding research activities.

The demise of 'intramuros' research activities concentrated within the research laboratories of large corporations and the implementation of open innovation systems that favour the outsourcing of the recombinant generation of technological knowledge to specialized knowledge-intensive businesses (KIBs), and academic departments might help the dissemination of technological knowledge.

Access to technological knowledge should be increased, favouring the distribution of universities and public research centres across the system so as to improve the proximity of firms to the available pools of public knowledge and reduce the distance of peripheral regions from the knowledge spillovers. In a similar vein the inflow of foreign direct investment (FDI) and the location of advanced multinational companies should be favoured as tools for local firms to access the spillovers of higher levels of technological competence.

The dissemination of existing technological knowledge should become the object of dedicated policy tools. The strengthening of relations between the business community and the public research system, specifically between firms and universities, might help the effective dissemination of knowledge and knowledge generating competence. Public policy should support all interactions between academics and firms, favouring the actual creation of additional pecuniary knowledge externalities with the provision of subsidies and fiscal allowances to all contracts between firms and the academic system. The dissemination and implementation of a fabric of good-quality public research centres and universities throughout the system is likely to generate better results than the concentration of centres of worldwide excellence in a few spots. For the same reasons, the mobility of skilled and creative scientists and experts among firms and between firms and research institutions at large can become the target of dedicated research policy interventions aimed at spreading competence and technological expertise.

CONCLUSION

This chapter has implemented an evolutionary approach that integrates strong Marshallian and Schumpeterian traits with the recent advances in the economics of complexity. Innovation can be considered as an emerging property of an economic system that takes place when its structural characteristics provide access to external knowledge as an indispensable input in the generation of new technological knowledge. Building upon the Marshallian legacy, external knowledge is considered an indispensable input, together with internal research activities, in the recombinant generation of new knowledge. The reappraisal of the Schumpeterian notion of innovation as a conditional result of a form of reaction to unexpected events led us to articulate the hypothesis that the reaction of myopic but creative agents that try to cope with the changing conditions of their product and factor markets may lead to the effective recombinant generation of new technological knowledge, and hence the actual introduction of productivity enhancing innovations when they are embedded in an organized complexity where they can actually take advantage of the external knowledge available within the innovation system in which they are embedded.

In this context ABM enabled us to explore the effects of alternative institutional, organizational and architectural configurations of the knowledge structure of the system in assessing chances to pursue effectively the recombinant generation of new technological knowledge and to introduce technological innovations. The introduction of innovations is analysed as the result of systemic interactions among learning agents. The reaction of agents may become creative, as opposed to adaptive, so as to lead to the introduction of productivity enhancing innovations when external knowledge can be accessed at low costs and used in the recombinant generation of new technological knowledge. Building upon agent-based simulation techniques, the chapter explored the roles that alternative configurations of intellectual property right regimes and architectural configurations of the system play in assessing these costs, and hence the chances to perform effectively the recombinant generation of new technological knowledge.

The results of the ABM confirm that a system characterized by high levels of knowledge dissemination is actually more effective in promoting the rates of introduction of technological innovations. The results, however, show that systems characterized by high levels of concentration could offer advantages in terms of faster discovery, due to the close relations that could be established among the knowledge producers. The implementation of ABM has enabled the rigorous framing of a complex system dynamics where innovation is the emerging property that takes place when a number of complementary conditions qualify the reaction of firms and make them creative. The simulation model can be applied to control the implications of an array of alternative settings and hypotheses concerning appropriability conditions, IPR regimes, knowledge generation routines and, most important, policy interventions that can alter the structure of knowledge flows so as to increase the levels of organization of the complexity of a system.

Taking inspiration from Schumpeter and Marshall, and recent developments in the analysis of the economic complexity of technological change, the ABM has shown the systemic conditions that make innovation possible. Innovation is an emergent property of the organized complexity of a system because innovation is the outcome of a situated and localized reaction when it can take advantage of a collective and situated process embedded in institutional as well as structural settings, and involving the combination of in-house and external knowledge and capabilities. Additionally, the uncertain outcome of these endeavours is portrayed as stochastic functions (lotteries), emphasizing that there is no automaticity as regards success in both activities. In this context an important aspect of the present elaboration is that the spatial distribution of innovation activities is explained endogenously from the interaction of competing firms.

Summarizing the results of the present simulation, it seems possible to assume that the more the knowledge producers such as universities and advanced science-based corporations are spread throughout a territory, the faster and more effective the innovation process becomes. Myopic but creative firms coping with the changing conditions of their product and factor markets are better able to improve their reaction and make it creative, as opposed to adaptive, when technological knowledge is disseminated in the regional, institutional and technological spaces.

NOTES

1. Schumpeter (1947: 149–50) makes the point very clear: 'What has not been adequately appreciated among theorists is the distinction between different kinds of reaction to changes in "condition". Whenever an economy or a sector of an economy adapts itself to a change in its data in the way that traditional theory describes, whenever, that is, an economy reacts to an increase in population by simply adding the new brains and hands to the working force in the existing employment, or an industry reacts to a protective duty by the expansion within its existing practice, we may speak of the development as an *adaptive response*. And whenever the economy or an industry or some firms in an industry do something else, something that is outside of the range of existing practice, we may speak of *creative response*. Creative response has at least three essential characteristics. First, from the standpoint of the observer who is in full possession of all relevant facts, it can always be understood *ex post*; but it can practically never be understood *ex ante*; that is to say, it cannot be predicted by applying the ordinary rules of inference from the pre-existing facts. This is why the "how" in what has been called the "mechanisms" must be investigated in each case. Secondly, creative response shapes the whole course of subsequent events and their "long-run" outcome. It is not true that both types of responses dominate only what the economist loves to call "transitions", leaving the ultimate outcome to be determined by the initial data. Creative response changes social and economic situations for good, or, to put it differently, it creates situations from which there is no bridge to those situations that might have emerged in the absence. This is why creative response is an essential element in the historical process; no deterministic credo avails against this. Thirdly, creative response – the frequency of its occurrence in a group, its intensity and success or failure – has obviously something, be that much or little, to do (a) with quality of the personnel available in a society, (b) with relative quality of personnel, that is, with quality available to a particular field of activity relative to the

quality available, at the same time, to others, and (c) with individual decisions, actions, and patterns of behavior.'

2. Empirical investigations and tests of specific hypotheses can complement and support agent-based simulations. See Antonelli and Scellato (2011) and Antonelli et al. (2011).

3. The microfoundations of evolutionary complexity: from the Marshallian search for equilibrium to Schumpeterian dynamics

Cristiano Antonelli and Gianluigi Ferraris

1. INTRODUCTION

The identification and appreciation of the Marshallian foundations of evolutionary thinking in economics is necessary to identify and overcome the limits of the biological evolutionary framework of analysis and to contribute to the new emerging evolutionary complexity with a consistent microeconomics of endogenous innovation that implements the reappraisal of the Schumpeterian notion of 'creative response' (Arthur, 2009; Kirman, 2016).

The Schumpeterian notion of creative response has received very little attention so far and yet is indispensable to go beyond the shortcomings of biological evolutionary approaches (Antonelli, 2015a, 2017b). Indeed, Schumpeter (1947: 150) provided a founding framework to grasping the endogenous microeconomic determinants of innovation. Innovation is the result of a creative response to unexpected events conditional to the availability of substantial externalities. Schumpeter defines the creative response as 'something that is outside of existing practice' by highlighting three essential characteristics: 'it cannot be predicted by applying the ordinary rules of inference from pre-existing facts. . . [it] shapes the whole course of subsequent events and their long-run outcome. . . its intensity and success or failure has . . . to do . . . with the socio-economic context.' The notion of creative response – and its contrast with the adaptive one – enables us to articulate an evolutionary complexity that puts the microeconomic determinants of the decision to innovate at the centre of the analysis. In so doing, reappraisal of the notion of creative response enables us to articulate an endogenous model of the innovation process that can be enriched and implemented by the explicit identification and appreciation of the Marshallian legacy (Antonelli, 2008, 2011, 2017a; Antonelli and Ferraris, 2011).

Specifically, this chapter articulates the view that analysis of the Marshallian and Schumpeterian microfoundations of endogenous innovation enables us to elaborate a clear analytical separation between the evolutionary approaches based upon biological metaphors and the new emerging evolutionary complexity overcoming the limits of the former and contributing to the latter. The integration of Marshallian notions of imitation externalities and selection processes with the Schumpeterian notion of creative as opposed to adaptive response provides a rich and coherent microeconomic analysis of the determinants of the innovation process at the firm level that is able to take into account the effects of the system in which the individual decision-making of heterogeneous agents takes place.

The rest of the chapter is structured as follows. Section 2 spells out the basic ingredients of the Marshallian and Schumpeterian frameworks of analysis, and shows how their integration provides coherent microfoundations for the dynamics of evolutionary complexity. Section 3 presents an agent-based simulation model (ABM) to test the coherence of the analysis. Section 4 explores the implications of the results. The conclusions summarize the main results of the chapter and confirm the central role of the creative response in the introduction of innovation based upon the use of knowledge externalities as the basic mechanism of an effective and microfounded evolutionary dynamics able to move away from the ambiguities of biological evolutionary approaches.

2. FROM THE MARSHALLIAN SEARCH FOR EQUILIBRIUM TO SCHUMPETERIAN DYNAMICS: THE BASIC ROLE OF EXTERNALITIES

In his essay in honour of Alfred Marshall, Joseph Schumpeter (1941) acknowledges the many contributions of the Marshallian legacy to his own understanding of the role of selective competition among heterogeneous firms. The Marshallian approach has been a fundamental and constant source of inspiration for Schumpeter, from *The Theory of Economic Development* (1911–34) to the 'The creative response in economic history' (1947).[1] The Marshallian approach, in fact, can be regarded as the foundation not only of these early contributions but also and primarily of Schumpeter's 1947 attempt to provide an endogenous understanding of the innovation process able to integrate the analysis at the firm level with the appreciation of the role of externalities embedded in the system.

The Marshallian model rests on three building blocks: i) exogenous innovations; ii) no appropriability; and iii) imitation externalities. Let us consider them in turn. In the Marshallian framework innovations are the starting point. For unknown reasons they are introduced occasionally and randomly. Their exogenous introduction puts – and keeps – the system in motion. According to Marshall, knowledge cannot be appropriated by inventors, and spills freely like information so that everybody is immediately aware of details of the best practice. Perfect access to the best knowledge at each point in time is a key aspect of the notion of 'normal' cost: 'But though everyone acts for himself, his knowledge of what others are doing is supposed to be generally sufficient to prevent him from taking a lower or paying a higher price than others are doing' (Marshall, 1920, V, 3, p. 199). As a matter of fact Marshall introduced the notions of limited appropriability and spillover well before Arrow (1962) and Griliches (1979). The imitation of exogenous innovations introduced randomly is the focus of the analysis and the engine of the dynamics both in Marshall and biological evolutionary models. Marshall considers two types of externality: agglomeration and imitation. Agglomeration externalities have received much attention, while imitation externalities have not. Yet imitation externalities are at the heart of the Marshallian dynamics that leads from variety to homogeneity by means of selection and from out-of-equilibrium to equilibrium. Marshall assumes that firms are heterogeneous: some firms perform better than others. Selective competition drives the system to generalize the competence of the best-performing firms. In Marshall, equilibrium is the result of a competitive process that reduces heterogeneity to homogeneity.[2] Exogenous and random innovations, and consequently the variety of firms, are the cause of the Marshallian imitation externalities. Externalities and variety decline together, along a competition process – intertwined with a selection process – that accounts for system growth but reduces variety and consequently destroys the very origin of externalities. They display their effects along with the selection process and the reduction of heterogeneity to homogeneity. Marshallian imitation externalities are endogenous to the system and intrinsic to the Marshallian search for equilibrium. As such, however, they are bounded.

Marshall assumes that a variety of firms try to produce, enter and exit the market place with different levels of productivity and costs. At each point in time firms are confronted with partial equilibrium that unveils their heterogeneity in terms of production costs. Less-efficient firms are sorted out while more efficient ones can enjoy the benefits of transient rents and increase their size. In the Marshallian process, new entrants and poorer-performing incumbents, however, can imitate freely the best-performing ones. The efficiency of best-performing firms spills freely in the system and can be accessed and shared by any other agent.[3] The imitative entry of new competitors and the imitation of incumbents affect the shifting position of the supply curve that engenders a sequence of lower market prices and larger quantities. The variance of profitability levels shrinks. In the long term the process leads to the eventual identification of the equilibrium price according to which only the most efficient firms can survive with normal profits. The identification of a stable equilibrium stops the endogenous generation of externalities. In equilibrium there is no growth. Growth lasts as long as the selection and imitation process that enables firms to push the allocation of inputs towards their most effective use is in place. Marshallian externalities are endogenous but bounded.

The influence of Marshall on *The Theory of Economic Development* is clear. The role of entrepreneurship is a first attempt to fill the Marshallian gap about the origin of innovations. Schumpeter however does not really provide an endogenous account of the origin and determinants of entrepreneurship. It remains unclear whether the flows of innovations introduced by entrepreneurs and their entry are steady through time and space, or exhibit relevant and systematic changes. As a matter of fact the evolutionary models that impinge upon Nelson and Winter (1982) are intrinsically Marshallian and are consistent with the legacy of Schumpeter (1911–34) where, following Marshall, innovations are exogenous as they are introduced by entrepreneurs that enter the economic system from outside without any economic causality, rather than the framework elaborated by Schumpeter in 1928 and 1942 and his great synthesis of 1947.[4]

The Marshallian dynamics of imitation externalities provides the foundations for the path-breaking contribution of Schumpeter (1947). This framework can be regarded as a fully fledged evolutionary process based upon the notion of endogenous innovation as the outcome of a creative reaction able to reshape the existing map of isoquants that takes place in out-of-equilibrium conditions when firms' plans do not meet the actual product and factor market conditions, provided the system is able to support their reaction with the provision of knowledge externalities.[5] If knowledge externalities are not available the response of firms will be adaptive and consist only in the traditional movements on the existing map of isoquants.

The Schumpeterian dynamics elaborated in the 1947 essay differs from the Marshallian one for two key reasons: first, in Schumpeter, externalities are knowledge externalities rather than imitation externalities. Knowledge externalities make it possible for every firm to introduce productivity-enhancing innovations that keep the system in a cost-reducing process further reinforced by the increased levels of generation of new technological knowledge that is able to reinforce the further creation of endogenous knowledge externalities. Second, in Schumpeter the creative reaction of firms supported by the self-sustained dynamics of knowledge externalities enables the introduction of innovations that are by definition the cause of unexpected changes in product and factor markets. Marshallian agents can imitate only from advanced firms. Advanced firms cannot take advantage of their transient competitive advantage to introduce new innovations. Schumpeterian agents, on the contrary, exhibit the distinctive features of entrepreneurship that enable them to react to both good and bad performance. In both cases, in fact, they will try to introduce innovations either to counter their decline and eventual exit or to take advantage of their competitive advantage and increase it with the introduction of new technologies. The levels of the actual reactivity of firms and of the quality of knowledge externalities provided by the system are the key variables of the Schumpeterian approach that enable firms to account for endogenous growth of output and productivity (Antonelli and Scellato, 2011; Erixon, 2016). Both are the results of implementation of the Marshallian framework. Identification of the Marshallian legacy enables us to better appreciate the strength of the late contribution by Schumpeter.

3.   THE SIMULATION[6]

The typical bottom-up approach of interactions nested in a systemic context of ABM provides an excellent tool for theoretical investigations. This use of ABM, besides its traditional application in forecasting, seems to open an innovative field of investigation to validate the robustness and consistency of theoretical hypotheses (Pyka and Fagiolo, 2007; Mueller and Pyka, 2016). ABM seems most appropriate to show how the implementation of a microfounded evolutionary complexity that integrates the legacies of Marshall and Schumpeter is able to overcome the microeconomic limits of the biological evolutionary framework with an endogenous account of the innovation process. By setting appropriate values for the key simulation's parameters (imitation externalities, knowledge externalities, knowledge governance and reactivity), the ABM, in fact, enables us to compare alternative bottom-up system dynamics: the Marshallian and the full range of Schumpeterian ones determined by the varying levels of reactivity and knowledge governance.

The ABM used to compute the simulations reproduces a stylized economy where a variety of firms produce a unique output, useful both for investment and consumption. The simulated economy is closed – neither import nor export activities are allowed – and systematically reach a state of local equilibrium: the whole production is sold, firms fully redistribute profits and shareholders immediately contribute to cover losses. No form of accumulation either in savings or equity is allowed. The levels of capitalization of firms are given and are maintained by the shareholders through immediate contributions to cover potential losses. Profits are always fully distributed, and do not eat/add capital funds of any kind.

Wages (w) are constant and there is an unlimited supply of labour. The cost of labour per unit is set by a simulation parameter. Provided that new technological knowledge is produced by employing labour, all the costs the firms afford produced a monetary transfer to the workers. The utility and demand functions of consumers, employees and shareholders are not simulated explicitly: their behaviour is summarized in the price equations of the goods. Output prices are set by a market maker that ensures that the entire, fixed, amount of money is allocated to consumption.

Because all remuneration (dividends, wages, contributions to research) will be used immediately to buy all the goods produced and money circulates twice for each production cycle, demand for goods, in value terms, is equal to the amount of money in the system. The velocity of circulation of money (per cycle) is equal to 2; all the money is paid in the form of remuneration of workers, researchers and shareholders (either as a positive amount in the case of profits or a negative one in the case of losses). During the same cycle all the money is spent to buy goods.

At the aggregate level the model can be summarized as:

$$Gs = +Y_i \qquad (3.1)$$

where Gs represents the aggregate supply, and Y are the individual revenues;

$$Gd = Gs \qquad (3.2)$$

where Gd represents the demand of goods, both for investment or consumption; and

$$Gp = M/Gs \qquad (3.3)$$

where Gp represents the price of a single unit of production and M the entire amount of money circulating in the system. Note that the entire amount of M is always available for consumption: when enterprises are subject to losses, the aggregate expense for salaries is greater than the value of production, so shareholders receive negative profits, and vice versa. At the aggregate level the amount of money available for buying products sticks always to M. The aggregate production function could be expressed as:

$$Y = (A)L \qquad (3.4)$$

where A is productivity and L the amount of labour the enterprise employs.

The productivity reflects both agglomeration effects and the purchase of knowledge as it may be influenced by the current amount of technological knowledge (T) that the firm is able to mobilize and by a small fraction (g) of the amount of technological knowledge mobilized in past production cycles, as follows:

$$A = \sum [L_i^* (a_i^* (1+ t_{i0} + g^* \sum t_{i-j})]/\sum L_i \qquad (3.5)$$

Equation (3.5) makes explicit the work of knowledge externalities: L_i represents the input the i-th enterprise employs; a_i its labour productivity; t_i the level of technological knowledge this enterprise has just bought; and $\sum t_{i-j}$ the sum of past technological knowledge acquisition, weighted by a decay parameter (g) – the fraction the enterprise permanently acquires in its knowledge estate for each technological knowledge acquisition. New technological knowledge acquisition is a risky activity, so its effects on productivity levels take place with a risk coefficient (R): each new technological knowledge acquisition (t_{i0}) could fail, and in that case the productivity of the single enterprise becomes:

$$A_i = (a_i^* (1 + g^* \sum t_{i-j}) \qquad (3.6)$$

Under Marshallian restrictions, the contribution of technological knowledge is zeroed. As a consequence productivity is limited and reaches a maximum that is given by the equilibrium level output in the system where all firms use the best technology. The productivity equation becomes:

$$A = \sum L_i^* a_i /\sum L_i. \qquad (3.7)$$

Each period a_i can be upgraded by a fraction of the difference between its level and the productivity level of the best enterprise, due to imitation effects. This upgrade is subject to risk, too, in this way:

$$A_{i1} = a_{i0} + h (a_{max} - a_{i0}) \mid E > = R \qquad (3.8)$$

where a_{max} is the productivity of the best in class enterprise; E is a random number tossed from a uniform distribution; and R measures the probability the imitation fails. When $E < R$ the productivity level of the enterprise remains at the latest reached level.

The cost equation includes the cost of both labour (w) and technological knowledge (z):

$$TC = wL + zT \qquad (3.9)$$

where L is the single input labour and w are unit wages; z is the cost of a technological knowledge unit; and T is the amount of employed technological knowledge.

Note that in equation (3.9) the effect of T on productivity is usually such that $\delta Y/\delta T = c > z$, where c is the cost of technological knowledge if it were a standard good, while z is the actual market costs of technological knowledge. There is an equilibrium level (c) of knowledge costs that reflects the equilibrium conditions for its generation. If technological knowledge were a standard economic good its cost, c, would be equal to w, the cost of labour. The case for adaptive reaction takes place when technological knowledge is acquired at its equilibrium cost, c. The use of T will allow firms to introduce novelties without direct economic effects on output levels that are higher than the total cost of the technological knowledge acquired. There is no chance for firms to introduce innovations that enhance output beyond the levels of the costs incurred to purchase the technological knowledge. When $z = c$ the value of the technological knowledge T matches the equilibrium value of its marginal product.[7]

If positive pecuniary knowledge externalities are at work the reaction applies successfully and becomes creative. This amounts to assuming that when the cost of technological knowledge falls below equilibrium levels ($z < c$), technological knowledge can be treated as a factor that enhances output beyond its costs. In this case, total factor productivity (TFP) increases because of the discrepancy between the equilibrium levels of technological knowledge costs and its actual (lower) levels. The economic effects of technological knowledge purchased at a cost z that is lower than the equilibrium cost c consist in the positive outcome of the reaction: the creative reaction takes place exactly in these circumstances. Firms can enjoy an 'unpaid' increase in productivity levels that is equal to the levels of pecuniary knowledge externalities – i.e. the difference between the equilibrium levels of the technological knowledge costs and their actual levels as determined by the working of pecuniary knowledge externalities. When, instead, the actual costs of knowledge are in equilibrium the reaction of firms will be adaptive.

The stylized economy configuration depends on a wide set of parameters. The model allows different set-ups to compare the simulation outcomes of different theoretical frameworks. For each simulation some parameters have a key role and vary to configure different simulation scenarios, whereas others are usually set at fixed values. The configuration of the economy used for the simulations was based on: i) the presence of 1000 agents; ii) the availability of 10,000 units of money for the whole transaction; iii) a fixed labour price (one unit of money) and an infinite labour offer; iv) out-of-business enterprises were replaced by new entrants with 20 per cent probability, whereas enterprises went out of business when their demand for factor became less than one unit. In order to set up an initial variety of agents, the employed labour and the productivity at the start of the simulation have been tossed randomly: labour was allowed to vary in the range]1,10[and productivity in the range]0.001,0.2[. Some agents have been endowed with higher values: respectively labour was tossed in the range]10,20[and productivity in the range]0.2,1.0[; the number of such 'smarter' agents was set to 15 per cent of the 1000 agents populating the economy. Agents were endowed with the capability to both adapt and react. Adaptation has been simulated by setting up the agents to increase or decrease the amount of the employed factor of 10 per cent if the previous production cycle ended with a profit or a loss. In computing their results (either profits or losses) agents rounded the amount with a tolerance of 0.001 due to computation matters.

Externalities was simulated as productivity enhancement subject to a failure risk of 10 per cent; technological knowledge was set as suitable for one production cycle only, except for a small fraction (0.1 per cent in the simulations) that is added to the knowledge estate of the firms to mimic the learning process that always takes place both at the workers level and organizational level during the innovation exploitation. In fact the contribution of new technological knowledge is immediately subject to a decay of 99.9 per cent (the parameter is named 'techDecay'). The Marshallian imitation externalities have been simulated simply by granting each cycle each agent a labour productivity increase of 1 per cent (the value is set by the parameter 'imitation'), the difference between their own productivity and that of the 'smartest' agents – i.e. the firms with the highest productivity in the whole simulated economy. As mentioned this upgrade may fail: the probability of success was set at 90 per cent. The Schumpeterian scenario was based on the possibility for the agents to buy technological knowledge instead of receiving labour productivity upgrades – the parameter 'imitation' was set to zero. The amount of technological knowledge each agent buys in each production cycle depended on two key parameters: 'techRate', which measures the reactivity of firms as a percentage of the total output each agent would invest; and 'governancePerformance', which measures the quality of the knowledge governance by means of the discount factor applied for the price of technological knowledge, as in equation 3.10:

$$z = \partial Y / \partial T * (1 - \text{governancePerformance}) \qquad (3.10)$$

Note that if governancePerformance is set to zero, the cost of technological knowledge becomes equal to its marginal contribution and no knowledge externalities are available in the system. In this way the behaviour of the agents cannot be reactive; they can only adapt their factor allocation. The computation of $\partial Y / \partial T$ has been based upon the production each agent is forecast to obtain at the end of the production cycle, in order to forecast the price the produced good will be sold at and set up a plausible base to compute the productivity of the technological knowledge in monetary terms to set up a plausible base to compute z.

The amount each agent invests is subject to financial constraints: because, intentionally, no financial institution has been included in the model, the whole investment amount has to be covered either by profits or through savings obtained by reducing the input of factors. The amount an agent invests can be expressed as:

$$I = \min (Y_{-1} * \text{techRate, profit}) \mid \text{profit} > 0 \qquad (3.11)$$

or as

$$I = \min (Y_{-1} * \text{techRate, labour}_{-1} *0,1*w) \mid \text{profit} < 0 \qquad (3.12)$$

where I represents the investment the i-th firm is going to afford in the production cycle 0; labour and Y stand, respectively, for the labour the i-th agent employed in the previous production cycle and the output of the i-th agent in the previous production cycle. Schumpeterian agents are credited with the capability to react to out-of-equilibrium conditions: they invest and purchase new knowledge both when they enjoy extra profits (equation 3.11) or face losses (equation 3.12).[8]

Table 3.1 exhibits the six simulation scenarios generated by alternative values of the three key parameters: imitation, techRate and governance Performance. The single scenario devoted to Marshallian externalities is the first one, named 'Imitation'; the second, named 'Zero Gov', was run to demonstrate that the results achieved with no knowledge governance are close to those of the Marshallian scenario. The others explore the

Table 3.1   Simulation scenarios

Scenario	Parameter values		
	Imitation	TechRate	GovernancePerformance
Imitation	0.01	0.00	0.00
Zero Gov	0.00	0.05	0.00
Low Gov Low React	0.00	0.01	0.01
Low Gov High React	0.00	0.05	0.01
High Gov Low React	0.00	0.01	0.95
High Gov High React	0.00	0.05	0.95

different combinations between high/low reactivity (techRate) and good/poor knowledge governance (GovernancePerformance). The parameter 'techLife' – which indicates how many production cycles, after the current one, the benefits of a technological knowledge acquisition lasts – was always set to 0; and 'techDecay' – which measures the fraction of the acquired technological knowledge that is lost after its usage – was always set to 0.999. Practically each knowledge acquisition gave a productivity advantage only for one production cycle, with the exception of the 0.1 per cent of the acquired technological knowledge that became a consolidated component of the knowledge of the enterprise.

The simulation process consists in repeating a sequence of actions, managed through a precise schedule, to control the information level of the agents and compute some statistics and aggregate figures. Before starting the simulations the model is charged to:

- create the planned number of agents and assign each of them different size and productivity;
- create a random generator for each agent and assign one to each in order to avoid indirect and uncontrolled influence among agents;
- create the market maker object that will manage exchanges and set prices for goods, factor and knowledge;
- initialize common variables called theWatch used by agents and marketMaker to synchronize their actions.

Each simulation cycle mimics a whole production cycle as shown in a flow chart (not included here) that illustrates the sequence of orders the model gives the agents and other components each simulation cycle. To avoid single pseudo-random distribution that could pollute the results, a large number of simulations have been run for each scenario, and results summarized as average values, paying attention to variance that was not significant. The evolution of the simulated economy has been studied by means of the trend of the global output and productivity and by computing concentration indexes: i) sum of the relative contribution to the global output by the three larger firms (GB method); ii) sum of the contribution of the four larger firms (US method); and iii) Hall and Tideman aggregation index.

Tables 3.2–3.4 summarize the average results obtained during 100 simulations, 2000 production cycles long, for each scenario.

All the simulations were executed under the same parameter configuration but with different random distribution; the random seed was set randomly for each run. The first evidence that comes out of the figures is that the results are not dependent on the random distributions, due to the fact

Table 3.2 Productivity and output obtained during simulations of the different scenarios

Scenario	Productivity				Output	
	T_1	Var	T_{2000}	Var	T_1	T_{2000}
Imitation	0.254332	0.000100	0.994559	0.000022	1868	9946
Zero Gov	0.252382	0.000110	0.833257	0.000229	1858	9845
Low Gov Low React	0.253934	0.000136	0.798269	0.000065	1871	10,010
Low Gov High React	0.254936	0.000139	0.864955	0.000188	1877	10,229
High Gov Low React	0.252215	0.000143	1.261606	0.004393	1851	15,654
High Gov High React	0.254803	0.000150	7.177898	0.807940	1880	84,908

Table 3.3 Concentration

Scenario	Concentration					
	H & T − T1	H & T − T_{2000}	CR-USA − T1	CR-USA − T_{2000}	CR-GB − T1	CR-GB − T_{2000}
Imitation	0.003292	0.001845	0.038695	0.212997	0.029390	0.210496
Zero Gov	0.003279	0.044264	0.038486	0.941844	0.029220	0.938796
Low Gov Low React	0.003297	0.047324	0.038533	0.944353	0.029270	0.943503
Low Gov High React	0.003281	0.043585	0.038362	0.940119	0.029123	0.934742
High Gov Low React	0.003280	0.045686	0.038758	0.942979	0.029453	0.942099
High Gov High React	0.003281	0.036878	0.038203	0.926782	0.029004	0.890148

the variance figures are negligible and confirm the results are robust and due to the endogenous dynamic the model is able to mimic.

Productivity and output values show that the Marshallian scenarios were systematically able to reach the highest productivity value the system was initially endowed with. In each of the 100 simulations, under Marshallian dynamics, the economy reached a stable equilibrium. The results of the Schumpeterian scenarios demonstrate the role played by knowledge governance: with no governance at all or with poor governance,

Table 3.4   *Correlation between productivity and concentration and the*
              three key parameters

Scenario	Correlation – productivity			Correlation – concentration		
	Imitation	TechRate	Governance-Performance	Imitation	TechRate	Governance-Performance
Imitation	0.223784	n.a.	n.a.	−0.707532	n.a.	n.a.
High Gov High React	n.a.	0.806627	0.277358	n.a.	−0.203093	−0.335518

either with high or low reactivity, the economy had a slow growth, achieving final figures close to those obtained under Marshallian simulation. No imitation was allowed in the Schumpeterian scenarios. Thus results confirm the importance of the quality of knowledge externalities. When knowledge governance is effective, externalities arise and both total output and productivity reach higher levels. The compound effect of good knowledge governance and high reactivity is able to push the system to overcome initial limitations by achieving productivity that is seven times higher than the highest level of the best firm, as endowed at the start of the simulation. As Tables 3.3 and 3.4 and Figure 3.1 confirm, strong knowledge externalities and high levels of reactivity lead not only to fast growth, but also to higher concentration, configuring product markets with a few large firms and many small competitors. The results of the High Gov High React scenario show that more than 90 per cent of the output was done by four firms (CR-USA value) and the three big ones cover 89 per cent of the production (CR-GB value), providing they were able to spend a larger amount in investing to buy technological knowledge. Their higher productivity parallels a strong increase in total output.

In the Schumpeterian scenario the system is always in evolution, with a growth that, in the presented case, is 10 times larger with respect to the results obtained by the Marshallian hypothesis, where the output reached a stable level when the system went to equilibrium. The Schumpeterian dynamic implies the impossibility of reaching a stable equilibrium, with cycles of growth and decline and a total output that oscillates with a trend to grow. The interpolation of the output level could be represented by a simple linear function (the straight line called 'Linear' in Figure 3.1) in equation (3.13); the interpolation is quite good, the r-square value = 0.9005:

$$8146.400 + 32.881x \tag{3.13}$$

Figure 3.2 shows the output trend between the 500th and the 1000th cycles, in order to make easier the catching of the cycles: cycles where periods of

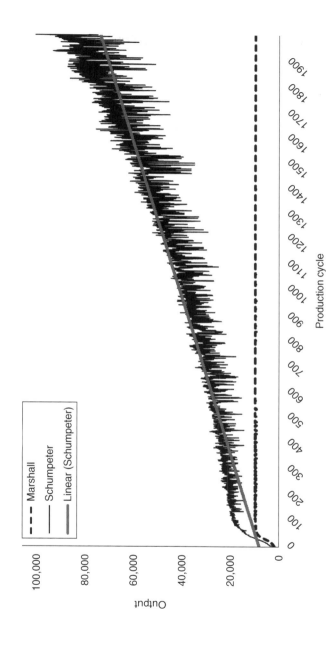

Figure 3.1 Comparison of production trend in Marshallian and Schumpeterian scenarios

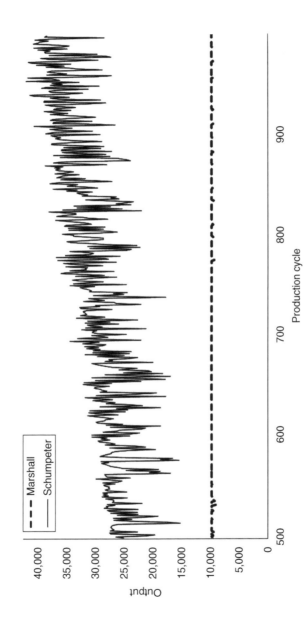

Figure 3.2 Comparison of production trend in Marshallian and Schumpeterian scenarios: zooming cycles between 500 and 1000

decay and growth alternate are clearly visible. In the Marshallian scenario the growth of output and productivity is larger the greater the variance of the given distribution of firms: growth is exogenous and bound to the maximum given level of productivity at the onset of the process. The variance of the distribution of firms, at the onset, instead, has no impact on the Schumpeterian scenario: growth is endogenous.

The robustness of the results was investigated by computing correlation indexes between key parameters and results. In order to exclude that the tolerance value, used only for computational matter, would affect the results a correlation was computed between its values and the corresponding outputs. The named indexes were computed using the results obtained by dedicated batches of simulations, where the parameters under investigation were randomly tossed in each simulation and the others – including random distributions, with the exception of the distribution used to toss the parameter under investigation – were fixed in order to isolate the effect of each single investigated parameter. Table 3.4 reports the values of the correlation between the key parameters and output values.

Table 3.4 shows the weak correlation between imitation and the levels of productivity achieved at the end of the simulations (the weakness of the correlation is witnessed by the variance of the achieved output level, which was 0.0038 compared with an average value of 9822.104, even if the values for the parameter imitation were tossed in the range]0,0.5[). Imitation level matters for concentration indexes because it determines the speed for reaching equilibrium and the time better firms have to grow due to their adaptive reaction (highly productive enterprises make profits and each time upgrade their input by increasing it, so their production grows and the system becomes more and more concentrated). Under the Marshallian legacy enterprises are doomed to reach the best productivity due to the work of imitation, so the intensity of the phenomenon has influence only on the pattern the system follows to reach the results, not on the final result. As expected, TechRate and GovernancePerformance exert a positive and significant impact on the Schumpeterian-based simulations.

Table 3.5 shows that the 'tolerance' parameter does not affect the results

Table 3.5   Correlation between the 'tolerance' parameter and results of the simulations

Scenario	Correlation – tolerance				
	Output	Productivity	H & T	CR-USA	CR-GB
High Gov High React	0.045423	0.015503	0.040802	0.072391	0.001194

of the simulations: the correlation between its value and the results has been computed through 100 simulations with configuration High Gov High React and tolerance values randomly tossed between 0 and 0.0011. The correlation values are meaningless and confirm that even with smaller or even null values for tolerance the results of the simulations are the same.

No other correlations have been investigated as this chapter aims to highlight the differences in results stemming from alternative theoretical configurations rather than forecasting values: the results of the simulation support the hypotheses outlined without any need for empirical evidence about the parameters configuration.

4.   ECONOMIC INTERPRETATION OF THE RESULTS OF THE SIMULATION

The results obtained running the simulation model under various configurations (scenarios) and different random values for the key parameters demonstrated that the dynamics theorized by Marshall and Schumpeter is fully reproduced by the model. The results confirm the strong and effective role of the microeconomic foundations of the innovation process. The stronger the reactivity of firms and the more effective the role of knowledge externalities, the more dynamic the system is. The response of Schumpeterian innovative entrepreneurs to the unexpected mismatches in the product and factor markets supported by a rich environment enables them to secure the provision of high-quality knowledge externalities accounts for economic growth. Reactivity and knowledge governance are clearly the key parameters that enable the Schumpeterian scenario to outperform the Marshallian. These results of the ABM confirm the hypotheses and validate at the same time the proximity and yet discontinuity of the Marshallian and the Schumpeterian analytical frameworks. The introduction of reactivity conditional to the quality of the environment is the element that distinguishes and qualifies the Schumpeterian approach from the Marshallian one. The results have proven to be robust because the described effects emerged independently of different values of innovation risks and possibilities that new firms can enter the market.

ABM enables us to explore the working of the system of interactions, transactions and feedback between individual actions and the structure of the system that qualify the simple but articulated Marshallian and Schumpeterian frameworks outlined in Section 3. ABM provides the opportunity to grasp the dynamics of competitive interactions among heterogeneous agents that, because of the working of endogenous externalities, are able to affect the structure of the system itself. This approach is actually able to model in a parsimonious and simple way the intrinsic complexity of the nested interactions among agents and the endogenous changes in the structure of the system that lie at the heart of both the Marshallian and the Schumpeterian frameworks (Mueller and Pyka, 2016).

ABM operationalizes, through the interactions among a large number of agents of our systems, the comparison between the working of a typical complex process characterized by the key role of Marshallian externalities and the Schumpeterian notion of creative reaction conditional to the actual availability of knowledge externalities (Schumpeter, 1947), enriched by the explicit assumption that the action of agents affects the structure of the environment, including the actual amount of pecuniary knowledge externalities (Lane, 2002; Lane et al., 2009). In so doing, ABM enables us to identify and highlight the complementarities, the sequential implementations and the theoretical differences between the Marshallian and the Schumpeterian frameworks. The Marshallian framework can be regarded as the case of zero reactivity and constrained externalities. The Schumpeterian framework starts as soon as reactivity is greater than zero and externalities affect the possibility to innovate.

The results of the Marshallian scenario show the high levels of firms' output elasticity to distribution of productivity. The larger the initial variance, the larger is output growth. The results of the replicator analysis are fully confirmed. The results of the Marshallian scenario also confirm that: i) growth is driven by exogenous assumptions about the distribution of the heterogeneity of firms; ii) growth is bounded to the productivity of the best-performing firm. When all (surviving) firms reach it, growth stops. Growth can be resumed only by tossing new innovators with higher levels of productivity. Only the introduction of exogenous innovations enables the Marshallian scenario to generate further growth: this is exactly the basic engine of the biological evolutionary approach.[9] The automatic introduction of exogenous innovations can take place either by entrepreneurs, as suggested by Marshall (1890/1920) and Schumpeter (1911–34), or by the unexplained upgrading of the routines of corporations, as in Nelson and Winter (1982). The hypothesis of exogenous innovation as the single possibility to keep the dynamics of the system, maintained by the evolutionary literature, is far from the approach eventually sketched by Schumpeter in 1928, articulated in 1942 and fully elaborated in 1947.

The results of the Schumpeterian scenario, in fact, confirm that the dynamics of the system is fully endogenous: the larger is the reactivity of firms and the better the quality of knowledge externalities, the larger the output growth – growth that is endless, provided the system is able

Table 3.6   Ingredients of the different levels of Schumpeterian dynamics

	Low reactivity	High reactivity
Poor knowledge governance $z = c$	Quasi-Marshallian convergence to equilibrium	Slow Schumpeterian regimes
Good knowledge governance $z < c$	Slow Schumpeterian regimes	High-powered Schumpeterian dynamics

to regenerate high-quality knowledge externalities. Table 3.6 provides a synthesis of the different Schumpeterian combinations of high and low levels of reactivity and poor and good knowledge governance that the ABM allows us to explore.

The results of the simulation make it possible to compare the Schumpeterian simulations and to appreciate their substantial complementarity, highlighting the key innovations introduced by Schumpeter (1947, 1928) and their implementation (Antonelli, 2008, 2011) in the Marshallian framework.

The results of the simulation of the Marshallian model confirm five relevant issues:

1. Innovation is exogenous. Marshall, as much as Schumpeter, assumes that some firms and agents are occasionally and randomly able to introduce better technologies and more effective organizations.
2. Limited knowledge appropriability was well known by Marshall. All existing firms can imitate the superior technology. In Marshall, however, imitation can benefit only laggards. Marshallian externalities consist in imitation effects. Imitation externalities augment and accelerate the reduction of variety brought about by the parallel selection process. The increase of output stems from the selection and imitation processes with the exclusion of poorer-performing firms and the progressive convergence of all production units to the best practice. When variety is erased and all the firms share the best practice, there is no longer room for growth and the positive effects of imitation level off.
3. The Marshallian model is the first and most effective attempt to graft the Darwinian selection process in economics. Innovation is the exogenous source of variety. Variety is transient. Variety in fact exists at the onset of the process and is wiped out by the process itself. Growth is explained by the extent of variety and by the selection process itself. The larger is the variety at the onset of the process, the larger the rate of growth. Growth stems from the sorting process of the firms that

are less efficient and by the generalization of the best practice to all the surviving firms. The rate of growth declines together with the reduction of variety brought about by the selection process. Equilibrium, reduction of variety, exhaustion of endogenous externalities and the end of growth coincide. When the selection process is over, all firms are able to use the best practice, and the allocation of resources can no longer be improved. There is no endogenous mechanism in the Marshallian process by means of which variety can be reproduced.

4. The replicator dynamics introduced eventually both in biology and in economics is clearly at work in the Marshallian model (Foster and Metcalfe, 2012). The evolutionary applications of the Marshallian model however make clear its intrinsic limitations: in the replicator dynamics variety is exogenous.

5. Biological evolutionary economics impinges upon the Marshallian legacy much more than currently assumed; and, in any event, it exhibits stronger elements of continuity with the Marshallian framework than with the Schumpeterian approach.

In the scenarios that build upon Schumpeter (1947) innovation is the endogenous source of variety: it is path dependent as it may be continuously recreated by the endogenous dynamics of the system (Page, 2011). Innovation, and hence variety, is generated by the reaction of firms to the changing conditions of product and factor markets. Innovation and variety are the products of internal feedback supported by the working of pecuniary knowledge externalities. Firms caught in out-of-equilibrium conditions by unexpected factor and product markets try to cope with either extra profits or losses with the introduction of innovations. All firms, both advanced and laggard, try to innovate. In order to introduce innovations firms need knowledge externalities to generate technological knowledge and innovate. Their reaction will be adaptive when and where knowledge externalities are weak. The response of firms, instead, will be creative and successful so as to make the introduction of innovations possible when the cost of knowledge is below equilibrium levels. Endogenous variety is constantly reproduced by the dynamics of the creative response: firms in out-of-equilibrium conditions in fact dare to face the risks associated by means of the purchase, use and generation of technological knowledge to innovate. The creative reaction is persistent when and where knowledge externalities are reproduced and the best practice keeps moving ahead. The process rests upon the interaction between the entrepreneurial behaviour of agents in out-of-equilibrium conditions and the systemic process that is at the origin of pecuniary knowledge externalities. Growth keeps increasing along the typical Schumpeterian cycles explained by the successful introduction of new technologies until knowledge governance mechanisms support the levels of pecuniary knowledge externalities.

5.   CONCLUSIONS

Evolutionary models that impinge upon biological metaphors miss a consistent microeconomic analysis of the determinants of the introduction of innovations. Appreciation of the Marshallian framework clarifies the weaknesses of the evolutionary approaches that impinge upon biological metaphors: as a matter of fact they are more Marshallian than Schumpeterian. This enables us to mark a clear distinction between the two evolutionary approaches: those that impinge upon biological metaphors and miss a consistent analysis of endogenous innovation and the new emerging evolutionary complexity. The integration of the Marshallian legacy and the late Schumpeterian frameworks, in fact, provides solid microfoundations to an evolutionary complexity that is able to account for the endogenous introduction of innovations. The integration in a single framework of the dynamic interaction of reactive decision-making at the firm level and changing provision of externalities as determined by the evolving structure of the system are the cornerstones of the new evolutionary complexity.

Externalities play a central role both in the Marshallian and in the Schumpeterian analysis of the working of the economic system. In the Marshallian legacy, however, imitation externalities are the cause and the consequence of the search for a stable equilibrium. In the Schumpeterian legacy, in contrast (especially of 1947), knowledge externalities are the cause and the consequence of the creative response of firms in out-of-equilibrium conditions that makes possible the introduction of innovations, persistent growth and evolution. The differences and yet the continuity and actual complementarity between the notions of externality implemented by Marshall and the Schumpeterian notion of creative response conditional to the availability of knowledge externalities seem clear. In both externalities are endogenous to the system. In Marshall they are bounded as they are intrinsically tied to the – transient – heterogeneity and variety of firms. In Schumpeter, instead, they may be continuously reproduced by the creative response of firms, the introduction of innovations and the consequent dynamics of the system.

Schumpeter builds his dynamics upon the Marshallian foundations. While Schumpeter (1911/1934) praises Leon Walras, Schumpeter (1947, 1941 and 1928) relies on Alfred Marshall moving away from an *ex ante* equilibrium that precedes production to an *ex post* quasi-equilibrium that follows production and market exchange, showing that equilibrium is impossible with innovation. The early Schumpeter shares with Marshall the hypothesis that innovations are exogenous and random. Not surprisingly, biological evolutionary models impinge upon this line of analysis. A line of analysis that had been overcome not only by Schumpeter (1942) but also by Schumpeter (1947) which draws from Marshall two key mechanisms: i) the competition process as a selection process, and ii) the endogenous dynamics of externalities; but it implements radically the Marshallian framework with the notion of creative response conditional to the availability of endogenous knowledge externalities. The approach elaborated by Schumpeter (1928, 1947) is able not only to go beyond the limits of the unexplained origins of variety assumed by Marshall as the result of a Darwinistic grafting, but also to provide the foundations of an evolutionary complexity.

The understanding of the complementarities and linkages between the Marshallian and the Schumpeterian (1947) legacies enables us to take advantage of the rich Marshallian framework and to use it to implement and strengthen the foundations of the new emerging approach to innovation as a creative response and an emerging system property where both agent-based decision-making and the properties of the system in which firms are embedded. The Schumpeterian framework elaborated in 1947 shares the basic microeconomic tools laid down by Marshall but departs from them – as well as from the earlier approach elaborated in *Theory of Economic Development* – as soon as firms in out-of-equilibrium conditions are credited with the capability to try to react to the unexpected conditions of both product and factor markets by means of the endogenous introduction of innovations.

In the Schumpeterian dynamics the reaction function of the firms plays a central role. The performance of each company is considered in equilibrium when results for the year fall in a spherical neighbourhood of zero whose amplitude is parametrically determined by the researcher. Firms are in out-of-equilibrium conditions when their profits – either below or above 0 – are far from equilibrium. Firms that experience performance below equilibrium try to reduce the use of the current input and to purchase knowledge. Firms that enjoy profits try to increase both the amount of labour and that of knowledge: output and productivity increase together.

In the Schumpeterian model, pecuniary knowledge externalities substitute and actually augment the transient Marshallian externalities associated with the imitation and selection processes. Because of the Arrovian characteristics of knowledge as an economic good such as limited appropriability and exhaustibility, non-divisibility and hence cumulability, limited rivalry in use and the recombinant generation of new knowledge where existing knowledge is an indispensable input, the cost of knowledge may be actually lower than in equilibrium conditions. When the system properties are such that the costs of absorption and use of external knowledge are below equilibrium levels, the creative reaction can take place.[10] The Schumpeterian dynamics may be endless provided that knowledge governance mechanisms remain at work, and hence the costs of knowledge keep falling below equilibrium levels.[11]

The results of this chapter provide a platform that can be enriched further to contribute complexity economics implementing the notion of innovation and knowledge as emergent properties of an economic system based upon the recursive interaction between individual action and the structure of the system (see Antonelli and Ferraris, 2017a). The structure of the system makes available pecuniary knowledge externalities. The entrepreneurial attempts of agents (caught in out-of-equilibrium conditions) to introduce innovations have multiple effects:

- They bring new changes in product and factor markets altering the expected equilibrium conditions upon which agents elaborate their productive plans.
- They expand the generation of new additional technological knowledge modifying the structure of interactions and transaction.
- Hence, they change the actual amount of pecuniary knowledge externalities available in the system.
- They reproduce variety within the system and push ahead the technological frontier so as to keep improving best practice.

In the framework that builds upon Schumpeter (1928, 1947) innovation, knowledge, variety and growth are endogenous and stochastic. This Schumpeterian framework enables us to identify the forces that may limit its reproduction. At each point in time knowledge governance mechanisms, and hence the amount of pecuniary knowledge externalities, may decline with the consequent decline of the reproduction of variety, the rates of introduction of innovation and the rates of growth. The continual implementation of effective knowledge governance mechanisms is crucial to keep the working of knowledge externalities. This dynamics is clearly path dependent, as opposed to past dependent. When and if the system is no longer able to support the regeneration of pecuniary knowledge externalities, the innovation process stops and the system identifies a stable attractor. The recursive interaction between individual decision-making and the system properties enable us to identify the introduction of innovations as an emergent and contingent system property (Antonelli, 2008, 2011, 2013a, 2015a, 2017b).

Evolutionary complexity empowered by the solid microfoundations of endogenous innovation derived from the Marshallian and Schumpeterian legacies seems far better equipped than evolutionary economics impinging upon the Darwinian legacy to grasp the endogenous drivers of innovation and economic dynamics. Growth as a result of the endogenous and contingent reproduction of variety, made possible by the introduction of innovations, may be endless provided that appropriate knowledge governance mechanisms are implemented and supported. The recursive interaction between individual action and the evolution of the properties of the system paves the way to elaborate effective strategies at the firm level to make it actually persistent. The implications for policy are most important as it becomes clear that all interventions aimed at increasing the quality of knowledge governance mechanisms that favour the persistence of pecuniary knowledge externalities and consequently the reproduction of innovative variety are likely to enhance economic growth.

ABM is confirmed as a powerful theoretical tool as it enables us to show how different theoretical frames are the result of the recombination of analytical bricks drawn from pre-existing models that alter the working of the system and yield new theories. In this case the shifting role of externalities plays a crucial role.

NOTES

1. See the contributions of Stan Metcalfe (2007a and b, 2009a and b) that have highlighted the Marshallian foundations of the early Schumpeterian framework.
2. See Metcalfe (2007a: 10, quoting Marshall, 1920; internal citations omitted): 'In a famous passage Marshall claims that the tendency to variation is the chief source of progress. This telling phrase captures in a single step the deep evolutionary content of Marshall's thought but "What is meant by this?" The rest of the *Principles* make clear that variation and progress are connected by a variation cum selection dynamic, Marshall's principle of substitution in which more profitable firms prosper at the expense of weaker brethren. Outcomes are tested in the market so that "society substitutes one undertaker for another who is less efficient in proportion to his charges". Indeed, in introducing a discussion of profit in relation to business ability, Marshall is quite explicit that this principle of substitution is a "special and limited application of the law of "the survival of the fittest". Furthermore, innovation is inseparable from the competitive process. For the advantages of economic freedom "are never more strikingly manifest than when a business man endowed with genius is trying experiments, at his own risk, to see whether some new method or combination of old methods, will be more efficient than the old". The relation runs two ways and mutually reinforces the links between free competition and business experimentation.'
3. See Ravix (2012: 53, quoting Marshall, 1920): 'In Marshall, entry and exit appears in different contexts. For instance, economic change leads to the distinction between "those who open out new and improved methods of business, and those who follow beaten tracks".'
4. Careful reading of the celebrated notion of 'forest trees' introduced by Marshall (1920)

is useful to support the hypothesis that young trees are the carriers of innovations that account for the growth of the system and the continual reproduction of out-of-equilibrium conditions: 'We saw how these latter economies are liable to constant fluctuations so far as any particular house is concerned. An able man, assisted perhaps by some strokes of good fortune, gets a firm footing in the trade, he works hard and lives sparely, his own capital grows fast, and the credit that enables him to borrow more capital grows still faster; he collects around him subordinates of more than ordinary zeal and ability; as his business increases they rise with him, they trust him and he trusts them, each of them devotes himself with energy to just that work for which he is specially fitted, so that no high ability is wasted on easy work, and no difficult work is entrusted to unskillful hands. Corresponding to this steadily increasing economy of skill, the growth of his business brings with it similar economies of specialized machines and plant of all kinds; every improved process is quickly adopted and made the basis of further improvements; success brings credit and credit brings success; credit and success help to retain old customers and to bring new ones; the increase of his trade gives him great advantages in buying; his goods advertise one another, and thus diminish his difficulty in finding a vent for them. The increase in the scale of his business increases rapidly the advantages which he has over his competitors, and lowers the price at which he can afford to sell. This process may go on as long as his energy and enterprise, his inventive and organizing power retain their full strength and freshness, and so long as the risks which are inseparable from business do not cause him exceptional losses; and if it could endure for a hundred years, he and one or two others like him would divide between them the whole of that branch of industry in which he is engaged. The large scale of their production would put great economies within their reach; and provided they competed to their utmost with one another, the public would derive the chief benefit of these economies, and the price of the commodity would fall very low. (Book IV. XIII. 3). But here we may read a lesson from the young trees of the forest as they struggle upwards through the benumbing shade of their older rivals. Many succumb on the way, and a few only survive; those few become stronger with every year, they get a larger share of light and air with every increase of their height, and at last in their turn they tower above their neighbours, and seem as though they would grow on for ever, and for ever become stronger as they grow. But they do not. One tree will last longer in full vigour and attain a greater size than another; but sooner or later age tells on them all. Though the taller ones have a better access to light and air than their rivals, they gradually lose vitality; and one after another they give place to others, which, though of less material strength, have on their side the vigour of youth' (Book IV. XIII. 4). *The Theory of Economic Development* can now be read as the evident grafting of the Marshallian intuition about the role of entrepreneurs as the vehicles of innovation and growth.

5. As a matter of fact Schumpeter had already overcome the limits of the exogenous role of entrepreneurship, not only in the 1928 essay but also and more consistently in *Business Cycles* (1939), where the cause/effect relationship between the phases of the economic cycle and the flows of innovations is investigated in depth, at least at the aggregate level.

6. See Antonelli and Ferraris (2011, 2017a, b) for complementary specifications of the basic ABM presented here.

7. The recent advances in economics of knowledge enable us to substantiate the dynamics of knowledge costs. Technological knowledge as an economic good is characterized not only by limited appropriability, but also by non-exhaustibility and non-divisibility (Arrow, 1962, 1969). Technological knowledge, moreover, has the unique characteristic to be at the same time the output of a dedicated process and a necessary, indispensable input in the generation of new knowledge as well as in the production of all other goods (David, 1993). Finally, the generation of technological knowledge is a recombinant process characterized by the central role of the stock of existing knowledge, both internal and external, to each learning agent (Weitzman, 1996; Fleming, 2001; Sorenson et al., 2006). The understanding of the unique characteristics of technological knowledge

as an economic good and the features of its generation process enables us to better grasp the dynamics of knowledge externalities. Each learning agent – not only worst-performing firms but also most advanced ones – can actually benefit from the spillovers of the knowledge generation processes at work in the system (Griliches, 1979, 1984). The actual access conditions to knowledge generated at each point in time, and hence the mechanisms governing its dissemination, are crucial to make persistent the working of pecuniary knowledge externalities (Antonelli and Ferraris, 2011, 2017a, b).

8. Agents that face losses reduce their input by a parametrical amount that was set to 0,1 in all the simulations.

9. See note 8.

10. In the simulation model we allow the actual costs of knowledge (z) to assume different values so as to show the effects of different types of knowledge governance mechanisms. In the first simulation run z is slightly smaller than the cost of knowledge as if it were a standard good (c) to mimic a system with weak knowledge governance mechanisms at work. In the second simulation run z is assumed to be much smaller than c so as to consider the case of high-powered knowledge governance mechanisms.

11. As noted, knowledge equilibrium levels would take place when the cost of external knowledge equals the costs of a standard good or when absorption costs are so high that the total costs of effective use of external knowledge equals the costs of knowledge as a standard good.

4. External and internal knowledge in the knowledge generation function

Cristiano Antonelli and Alessandra Colombelli

1. INTRODUCTION

This chapter contributes the analysis of the process by means of which new technological knowledge is generated. We define the object of our analysis as the knowledge generation function, as distinct from the technology production function. The knowledge generation function frames the generation of new knowledge as an economic activity, while the technology production function analyses the contribution of technological knowledge to the production of other outputs. The knowledge generation function was introduced in the literature, somewhat accidentally, by Zvi Griliches (1979) and actually articulated by Adam Jaffe (1986), enriched by Richard Nelson (1982) and fully elaborated by Weitzman (1996, 1998), Fleming and Sorenson (2001) and Arthur (2009), who characterize the generation of new technological knowledge as the output of the recombination of existing knowledge.

The aim of the chapter is to implement the notion of the knowledge generation process as a recombination process, highlighting the complementary – as opposed to supplementary – indispensable role of the knowledge generated by other firms as a strictly necessary input, together with the size and composition of each firm's stock of both internal and external existing knowledge and their current research and development R&D activities.

The rest of the chapter is structured as follows. Section 2 provides a synthesized account of the emergence of the knowledge generation function and the increasing attention to the complementary – as opposed to supplementary – role of external knowledge and to the specific characteristis of the stock of knowledge both internal and external in each firm, instead of considering it a generic bundle. Section 3 articulates the hypotheses about the complementarity of external knowledge and the role of the composition of the stock of knowledge with R&D activities in the knowledge generation process. A novel specification of the knowledge generation function is the result of the analysis. Sections 4 and 5 present the data, the econometric model and the methodology used to test the hypotheses presented in Section 3. Section 6 presents the results of the econometric tests. The conclusions summarize the results of the analysis and explore the implications.

2. FROM THE TECHNOLOGY PRODUCTION FUNCTION TO THE KNOWLEDGE GENERATION FUNCTION

The notion of knowledge generation function studies specifically the process and inputs that make possible the generation of knowledge as an output.[1] The knowledge generation function differs from the technology production function, one of the pillars of the applied economics of innovation, where knowledge, alongside capital and labour, is considered explicitly as an input in the standard production function for all other goods (Griliches, 1979, 1990, 1992; Romer, 1990; Link and Siegel, 2007; Cohen, 2010).

The identification of the knowledge generation function is the result of a long process that takes its first steps from the Arrovian analysis of knowledge as an economic good (Arrow, 1962, 1969; Nelson, 1959). Preliminary attempts to identify a direct relationship between inputs and outputs in the generation of technological knowledge were first made in the growth literature by Phelps (1966), who called it the 'technology function' and 'effective research function', and by Gomulka (1970), who referred to the direct relationship between knowledge inputs and knowledge outputs as the 'production function of innovations'.

Griliches (1979) explores the relationship between R&D and knowledge output as a by-product of his attention to the structure of lags of R&D that is necessary to compute properly the stock of R&D capital, and mentions the 'complication' that knowledge is itself a dependent variable in a knowledge production function where past and current R&D efforts are the independent variables.[2] It seems worth noting that Griliches, who introduced the technology production function and the notion of spillover to highlight the role of external knowledge alongside internal knowledge in the production of all other goods, never considered the role of external knowledge in the generation of new technological knowledge, limiting its role to the production of all other goods. At the same time, Griliches should be credited with the first explicit appreciation of the role of the stock of knowledge together with that of current R&D efforts: in so doing he acknowledges the role of knowledge cumulability, but not of knowledge externalities.

Jaffe (1986) can be regarded as the first explicit, empirical analysis of the knowledge generation function: the number of patents is treated as the output of a Cobb–Douglas production function where internal R&D expenditures together with the spillovers of related knowledge generated by other firms are the independent variables.[3] As a matter of fact little attention has been paid in the subsequent literature to the Cobb–Douglas specification of the knowledge generation function used by Jaffe. The Cobb–Douglas specification proposed by Jaffe makes clear that external knowledge is not a supplementary input that magnifies the output, but an indispensable one. The Cobb–Douglas specification used by Jaffe implies that external knowledge is a complementary – as opposed to supplementary – input that cannot be dismissed: in a region with no spillovers available all internal R&D efforts cannot generate any technological knowledge. Very little attention has been paid to the complementarity of external knowledge as an indispensable input: the literature has mainly treated it as a supplementary one.

Attention on the efficiency of the knowledge generation function has been stressed by Nelson (1982), who made an important contribution to articulate a clear knowledge generation function approach, stressing the need to consider explicitly knowledge as the output of a specific and dedicated activity and to take into account the variety of inputs, complementary to R&D expenditures, that make possible the generation of new knowledge. David (1993) marks the final step in the process, stressing that knowledge is at the same time an input in the generation of new knowledge and in the production of all other goods; but it is also and primarily the output of a dedicated activity.

Weitzman (1996) opened up a new literature that frames the new understanding of knowledge generation as a recombinant process where existing knowledge items enter as inputs, shedding new light on the role of knowledge indivisibility and cumulability. After much attention was paid to the notion of knowledge non-appropriability, other key characteristics – such as non-exhaustibility, cumulability and complementarity that stem from its intrinsic indivisibility – are appreciated. Because no agent can command all existing knowledge, the recombination approach clearly implies a multiplicative relationship both between knowledge items and, at the firm level, between internal and external knowledge.

Antonelli (1999) articulates the hypothesis that technological knowledge is the output of a bundle of dedicated activities such as learning, R&D, search and technological transactions and interactions with the scientific community that enable firms to acquire the four knowledge inputs – that is, internal, external, tacit and codified knowledge – that are necessary for the generation of new technological knowledge. Along these lines,

Patrucco (2008, 2009) uses explicitly a Cobb–Douglas specification of the knowledge generation function to stress at the same time the complementarity between external and internal knowledge and their substitutability within well-defined ranges. Fleming and Sorenson (2001), Lucas (2008) and Arthur (2009) show how the generation of technological knowledge follows a branching process where earlier modules feed the new ones. Da Silva (2014) is able to model the 'Jefferson's candle light effect' with the notion of knowledge multiplier defined by the extent to which external knowledge enhances the innovative capacity of each firm. The larger the knowledge multiplier, the stronger the cumulative positive effects of external knowledge on the generation of new knowledge.

The specification of knowledge generation as a combinatorial process where, together with levels of current R&D efforts, the stock of existing knowledge both internal and external to each agent is an essential input in the current generation of new knowledge can be regarded as the arrival point of a long research process that has moved external knowledge from the role of supplementary input to the crucial role of complementary input.

Since then a growing empirical literature has paralleled the advances of the theoretical analysis exploring the characteristics of the knowledge generation function, assessing the role of different measures and proxies for both knowledge inputs and knowledge outputs, providing important contributions to understanding the complementarity of external knowledge to internal research and learning processes and the heterogeneity of knowledge as a composite bundle of differentiated items.

The first empirical estimates of the knowledge generation function are quite simplistic and use R&D expenditures as the key input. Furman et al. (2002) follow Griliches, and show that the differences in the levels of innovation activity across countries are explained by the differences in the level of inputs such as R&D manpower and spending invested in the generation of innovations. The empirical framework of the knowledge generation functions has been applied not only at the country level but also at the regional level, with interesting results with an elementary frame where R&D expenditure is the single input and the patents granted to a given region are the output (Fritsch, 2002; Ó hUallacháin and Leslie, 2007).

The contribution by Crépon, Duguet and Mairesse (1998, hereafter CDM) marks a major step in the empirical analysis from many viewpoints. To start with, the CDM approach provides the first econometric analysis of the knowledge generation function combined with the technology production function in a single framework. In so doing it provides a broad econometric model in which the relationships between knowledge inputs, knowledge outputs and productivity are estimated within a four equation model able to assess in parallel the effects of R&D expenditures not only on innovation counts but also on labour productivity and total factor productivity (TFP). In this sense the CDM model represents an empirical setting dedicated to the analysis of the relationship between innovation and productivity able to take into account also the generation of technological knowledge (Mohnen and Hall, 2013). The CDM approach, however (following strictly Griliches, 1979, 1990), does not consider the role of external knowledge in the generation of new technological knowledge. As a matter of fact the structured system of equations of the CDM approach raises the problematic issue of assigning knowledge externalities: whether they enter the technology production function and/or the knowledge generation function.

Parisi et al. (1996) explore the likelihood that R&D expenditures affect the introduction of product innovations, as distinct from process innovations, using the European Community Innovation Survey. In so doing they depart from the simplistic knowledge generation function and draw attention to the need to explore other determinants of process innovations.

Nesta and Saviotti (2005) make a major innovation in the empirical analysis of the determinants of the generation of new knowledge at the firm level, overcoming the limitations of regarding knowledge as a homogeneous bundle and drawing attention to the characteristics of the stock of knowledge as a central input in the generation of new knowledge and analysing the relationship between the coherence of the knowledge base and the innovative performance of US pharmaceutical firms measured by means of citation-weighted patent count.

The composition of the stock of knowledge, in other words, matters as well as its sheer size. The shorter the distance between different types of knowledge, the higher the capability of a firm to absorb external knowledge and to combine it with its internal knowledge, cumulated over time. Furthermore, this representation provides the basis to move to empirical analyses by constructing an image of the knowledge base as a network in which the nodes are constituted by units of knowledge at a given level of aggregation. Such empirical investigations can be successfully conducted by exploiting information contained in patent documents exploiting as a relevant source of information on the qualification of each knowledge item the assignment of patents to multiple technological classes (Saviotti, 2004, 2007; Krafft et al., 2009; Nesta and Dibiaggio, 2003; Quatraro, 2010; Colombelli et al., 2013a, 2014).

Conte and Vivarelli (2005) identify formal R&D activities and the acquisition of external knowledge as the key inputs of three alternative measures of innovative output. Their evidence confirms that external knowledge is a necessary and indispensable input strictly complementary to internal R&D activities.

Because of the relevant costs of absorption of external knowledge, knowledge externalities can be considered as pecuniary rather than pure externalities. Knowledge spills, but its use – as an input in the generation of new knowledge – is not free; dedicated resources are necessary to use external knowledge as such knowledge externalities are pecuniary rather than pure (Cohen and Levinthal, 1990; Colombelli and von Tunzelmann, 2011). Following this line of analysis, Gehringer (2011a, b, 2012) uses input–output tables to track the flows of pecuniary knowledge externalities and provides substantial evidence on the key role of external knowledge provided by upstream suppliers to downstream users, showing how the flows of knowledge parallel the flows of goods contributing to the downstream generation of new technological knowledge. Her evidence confirms, once more, the relevance of user–producer interactions (see also Von Hippel, 1988).

Jaffe (1989) applies his original approach to explore the effects of academic research on the generation of technological knowledge by firms localized in proximity. He is able to identify significant effects of academic research on the number of patents filed by firms particularly in the areas of drugs and medical technology, and electronics, optics and nuclear technology. Once again, Jaffe made one step forward, suggesting that academic research may actually induce R&D activities of co-localized firms.

Strumsky et al. (2012) and Youn et al. (2015) provide rich evidence, based upon US patent records dating from 1790 to 2010, to analyse the evolving distribution of patents across technological classes. They also explore the characteristics of the combinatorial process that is at the heart of the knowledge generation process, identifying the central role of the mix of exploitation and exploration of the existing technological knowledge.

An array of detailed empirical studies shows that external knowledge can be sourced by means of a variety of tools such as user–producer interactions, mobility of personnel among firms and interactions with academic research. In all cases the specific screening and reassessment of the items of existing knowledge are necessary for their subsequent active inclusion and integration as inputs of the knowledge generation process. The current levels of R&D expenditures of each agent do play a role, but only in a context that is shaped by the past of each firm and by the characteristics of the system in which it is localized (Feldman, 2003; Chesbrough et al., 2006; Antonelli et al., 2013).

Love and Roper (2009) explore the use of external knowledge in four discrete stages of the innovation process – new product development (identifying new products), product design and development, product engineering and product marketing – and test their complementarity with internal research activities in an innovation production function where output is measured by the percentage of sales derived from innovative products. Their empirical results, however, are influenced by the methodological approach that does not allow taking into account the positive effects of the substitution of any of the activities with others according to changes in their relative costs.

Johansson and Lööf (2008, 2014, 2015), in a remarkable string of contributions, draw attention back to the role of spillovers to generate technological knowledge, and stress the mutual complementarity of internal effort and external knowledge. While much attention is paid to the role of spillover, little attention is paid to the need for agents to participate actively in the process. Internal efforts are necessary both to scan the available external knowledge and to absorb it. Agents with low levels of commitment to R&D activities can benefit very little from the generous opportunities offered by regions rich in technological spillovers (Veugelers and Cassiman, 1999; Cassiman and Veugelers, 2006; Griffith et al., 2003; Ruckman, 2008).

3. THE HYPOTHESIS: EXTERNAL KNOWLEDGE AS AN ESSENTIAL INPUT IN THE RECOMBINANT GENERATION OF NEW TECHNOLOGICAL KNOWLEDGE

The recombinant knowledge approach has paved the way to elaborate a new frame of analysis able to accommodate the central role of existing knowledge, including external knowledge, as an input in the generation of new knowledge. As Weitzman recalls (1996: 209), 'when research is applied, new ideas arise out of existing ideas in some kind of cumulative interactive process that intuitively has a different feel from prospecting for petroleum'. This insight has led to the so-called recombinant approach: new ideas are generated by means of the recombination of existing ideas under the constraint of diminishing returns to scale in performing the R&D activities that are necessary to apply new ideas to economic activities. The generation of new knowledge stems from the search and identification of elements of knowledge that had been already generated for other purposes and yet reveals characteristics and properties that had not been previously considered. The search for existing knowledge items that can be recombined and used as input in the generation of new knowledge is strictly necessary.

Existing knowledge is both internal to the firm, stored in the stock of competence and knowledge accumulated in the past, and external to each firm. In this case it can be accessed by means of knowledge interactions and transactions with suppliers, customers and other agents qualified by substantial proximity. Appreciation of the generation of new technological knowledge as a recombination process that consists in the reorganization and reconfiguration of the relations among existing knowledge items enables us to better appreciate the effects of knowledge indivisibility, articulated in internal cumulability and external complementarity in the generation of new knowledge. The generation of new technological knowledge at each point in time, by each agent, in fact is strongly influenced not only by the internal accumulation of knowledge but also by the knowledge made available by the other firms that belong to the system in which each firm is embedded.

The recombinant approach enables us to appreciate the central role of two important inputs in the generation of new technological knowledge, such as the knowledge base of each firm as qualified by the size and the composition of stock of knowledge that each firm possess, and the components of knowledge that are external and yet highly complementary to the research activities undertaken by each firm.

Analysis of the composition of the internal stock of knowledge enables us to qualify the characteristics of the knowledge base of each firm. Technological knowledge cannot be regarded as a homogeneous pile, but rather as a composite bundle of highly differentiated and idiosyncratic elements that are qualified by specific relations of interdependence and interoperability. This approach enables us to identify the extent to which the generation of new technological knowledge in a field depends upon the contributions of knowledge inputs stemming from other fields: a new knowledge item exhibits high levels of composition when it relies upon a large number of other knowledge fields.

Our basic hypothesis is that external knowledge is an essential and strictly complementary – as opposed to supplementary – input in the generation of new knowledge. At each point in time, no agent possesses all the knowledge inputs. Yet agents need to access the variety of knowledge items that are available in the system and, as such, are possessed and being used by the other firms that belong to the system. The search for and the absorption of external knowledge is necessary, and its acquisition is the result of an intentional activity. Access to external knowledge is possible only if dedicated resources are invested to search, screen and interact to access, purchase, learn, understand and eventually recombine the external units of knowledge with the internal ones.

The standard Edgeworth definition of complementarity – according to which two activities are complementary if doing more of one activity increases the returns from doing the other – applies to two interdependent but separate activities, but not to the knowledge generation process where the mix of external and internal knowledge is the result of a constrained choice and includes the well-known possibility that, within the constraints of their relative costs, in order to increase output it may be necessary to increase the amount of external knowledge and reduce the amount of internal knowledge, or vice versa. An increase in pecuniary knowledge externalities stemming from the reduction of screening and absorption costs of external knowledge leads to the use of more external and less external knowledge, and yet to an increase in knowledge output.

The integration of these issues in the recombinant approach to the generation of technological knowledge enables us to lay down our basic argument. Technological knowledge is at the same time an input and an output, and it is the result of an intentional economic action. Technological knowledge is localized in the accumulated competence of firms and in the knowledge space in which firms are rooted. New knowledge can be generated, by means of the recombination of existing knowledge items, when, where and if:

- An intentional action directed at its generation is undertaken. New technological knowledge does not fall like manna from heaven. The generation of knowledge requires an active and explicit action. R&D activities are necessary to activate the recombination process. This is not their single function. Current R&D activities are also necessary to access, learn and absorb the stock of existing knowledge within the firm and the external knowledge generated by third parties. Current R&D activities are necessary also to track the records, retrieve and eventually reuse the knowledge that has been produced in the past and is stored in the stock of knowledge and competence that each firm has accumulated. Current R&D activities moreover are necessary to access, learn and absorb the stock of knowledge that is external to each firm: this contrasts the passive attitude that would characterize the prospective users of technological spillovers. In sum, current R&D expenditures are necessary not only to perform the recombination but also to supply it with the stock of knowledge internal to each firm and with access to the external knowledge generated by other firms.
- The knowledge base of each firm is identified and the role of previous knowledge is fully appreciated. The knowledge base of a firm is identified by the size and composition of the stock of knowledge that each firm has been able to generate in the past. The knowledge base exerts its positive effects in the long run and enters directly as an input in the knowledge generation function. According to our interpretative framework, knowledge generation is a non-ergodic

process where history matters as it helps build higher and higher levels of competence and innovative capability.

- External knowledge is a crucial, indispensable input in the generation of new technological knowledge. Each agent has access only to local knowledge interactions and externalities, that is, no agent knows what every other agent in the whole system at large knows. Because of the localized character of knowledge externalities and their strong tacit component, interactions matter. Consequently, proximity in a multidimensional space, in terms of distance between agents and their density, matters. Agents are localized within networks of transactions and interactions that are specific subsets of the broader array of knowledge externalities, interactions and transactions that take place in the system. The wider and easier the access to the local pools of knowledge, the larger the amount of technological knowledge that each firm is able to generate for given levels and composition of the internal stock of knowledge and the amount of current R&D.

- External and internal knowledge and R&D activities are mutually complementary. This has two important implications: i) no agent can generate new technological knowledge without access to external knowledge; ii) no agent can generate technological knowledge without appropriate efforts even in a spillover-rich context.

4. DATASETS

Our data source is the Intergovernmental Preparedness for Essential Records (IPER) database, which collects information on 3882 companies listed on the main European markets (UK, Germany, France, Italy and the Netherlands).[4] The IPER database has been built by matching information from multiple sources of data. Our main source of market and accounting data is Thomson Reuters Datastream, which delivers worldwide economic and financial time series data. To obtain additional relevant variables we include information collected from Bureau Van Dijk's Amadeus, which contains financial data on European companies. In order to match information from these two databases we use the International Securities Identification Number (ISIN), which uniquely identifies a security. Our final sample includes all companies listed on the selected markets and active in 2010 for which we were able to collect the necessary information from the above-mentioned data sources, and represents 42 per cent of the population of active listed companies in the period and countries analysed.

We also use data from the Organisation for Economic Co-operation and Development (OECD) REGPAT database, which provides regional information on the addresses of patent applicants and inventors as well as on technological classes cited in patents granted by the European Patent Office (EPO) and the World Intellectual Property Organization (WIPO), under the Patent Cooperation Treaty (PCT), from 1978 to 2006. In order to match the firm-level data with data on patents, we draw on the work of Thoma et al. (2010), which develops a method for harmonization and combination of large-scale patent and trademark datasets with other sources of data, through standardization of applicant and inventor names. Finally, we pooled the dataset by adding industry-level information on OECD countries from the organization's STAN database.

Our sample is an unbalanced panel that includes 1156 R&D reporting firms and 484 companies that submitted at least one patent application to the EPO in the period 1995–2006. Moreover, in order to build the cognitive distance index, which measures the dissimilarity between two technologies (see Section 5.1 for details), equation (4.4) focuses on 101 companies that submitted at least one patent application to the EPO only if this patent has been assigned to at least two technological classes.

This may cause some biases in the estimations if missing data are non-random. The econometric techniques we implement (which are described in detail in Section 5) allow for the mitigation of this problem. In particular, the estimation methods implemented allow us to account for the fact that only a subsample of firms are engaged in research activities, and that only relatively few firms have patents.

As mentioned above, the period of observation for all the firms examined is 1995 to 2006. Table 4.1 reports the sample distribution by macro-sector classes. High- and medium-high technology classes are highly represented in the sample of R&D reporting and patenting firms, while

Table 4.1   Sample distribution in macro-sectors, %

Macro-sector	Total	R&D reporting	Patenting
HT	16.98	29.31	32.75
MHT	17.31	27.93	37.91
MLT	3.18	3.39	4.53
LT	13.44	9.10	10.11
KIS	38.73	25.77	10.11
LKIS	6.95	0.78	0.91
EP	3.41	3.71	3.67
Total	100.00	100.00	100.00

knowledge-intensive services are highly represented in the whole sample and in the sample of R&D reporting firms.

5. METHODOLOGY AND VARIABLES

Our econometric strategy takes into account the many problems inherent to the model and to the nature of the data. First, only a minority of firms are engaged in (formal) R&D activities, so that studies restricted to these firms are prone to such biases. Also, only relatively few firms have patents, and thus analyses limited to them may be similarly biased. In addition, patents being count data require specific econometric methods to handle them. Finally, there is the major issue of the endogeneity of innovative input and output, and more generally of the simultaneity in our model as R&D is endogenous in the innovation equation.

We treat all these estimation problems by relying on a recursive model that is an extension of the CDM model developed by Crépon et al. (1998). In so doing, we rely on a model comprising a system of equations. We take care of selection problems and of the specific nature of variables by using a Heckman procedure and a count data specification for patents, respectively. We deal with simultaneity by using a two-stage estimation procedure.

It is worth noting that the original CDM model is based on three distinct interrelationships: i) the research equation linking the knowledge input with its determinants; ii) the knowledge generation equation relating knowledge input to knowledge output; iii) the productivity equation relating knowledge output to productivity.

Following this seminal work, more recently a number of studies have applied the CDM model to analyse the determinants of the knowledge generation process. In this stream of the literature it is possible to identify two groups of works: i) those studies based on an application of a fully structured CDM model, taking into account all three stages of the model; ii) those studies based on a partial structured model, taking into account at least one link between the three stages. Given our focus on the role of knowledge in the generation of new technological knowledge, we follow this latter group of works and apply an extended version of a CDM partial structure model.

More precisely, we focus on the first two stages of the CDM model and analyse the determinants of the knowledge inputs and the equations that test the knowledge generation function. First, firms decide on whether to perform R&D and, if so, how much. Then, depending on the extent of their R&D and other factors, they achieve a certain knowledge output. Hence, our model consists of two groups of equations:

- *Research and development equations:* 1) firms' decision to engage in R&D activities; 2) the determinants of the amount of R&D activities, and
- *Knowledge generation equations:* 3) the probability to patent; 4) the knowledge generation function.

The following subsections describe the econometric methodologies and the specifications used for the estimations of the model's four equations.

5.1   R&D Equations

To describe a firm's research behaviour, we rely on a two-equation model, where the first equation accounts for the fact that the firm is engaged in research activities and the second for the intensity of these activities.

Let $D\_R\&D_{it}^*$ be the latent dependent variable whether to invest in R&D or not, and $LnR\&D_{it}^*$ be the latent or true intensity of R&D investment of firm i; $D\_R\&D_{it}$ and $LnR\&D_{it}$ are the corresponding observed variables. The two-equation R&D investment model is written as follows:

$$D\_R\&D_{it}^* = \beta_1 x_{it}^1 + k_i^1 + u_{it}^1 \qquad (4.1)$$

with $D\_R\&D_{it} = 1$ if $D\_R\&D_{it}^* > 0$, $D\_R\&D_i = 0$ otherwise.

$$LnR\&D_{it}^* / (D\_R\&D_{it} = 1) = \beta_2 x_{it}^2 + k_i^2 + u_{it}^2 \qquad (4.2)$$

with $LnR\&D_{it} = LnR\&D_{it}^*$ if $LnR\&D_{it} > 0$, $LnR\&D_{it} = 0$ otherwise.

Here x_{it}^1 and x_{it}^2 are the explanatory variables; β_1 and β_2 are the respective coefficients; k_i^1 and k_i^2 represent unobserved characteristics that are fixed over time; and u_{it}^1 and u_{it}^2 are the individual-specific unobserved disturbances. The variable $LnR\&D_{it}$ is only observable if $D\_R\&D_{it} = 1$.

The independent variable explaining, first, the probability of engaging in R&D activities and, second, the intensity of these activities is intangible assets. Investment in intangible assets seems to provide a reliable proxy for the broad array of activities that are necessary to explore the existing stock of knowledge, both internal and external to each firm, master the recombinant generation of new technological knowledge and exploit it. Moreover, in the selection equation (4.1) we include a measure of firm size. Finally, both equations include a set of industry and time dummies to capture market and cycle conditions (see Section 5.3 for detailed specifications of all variables).

We estimate equations (4.1) and (4.2) following two approaches. The first approach is the Heckman two-step sample selection procedure (Heckman,

1981). Hence, the first equation is estimated using a probit; the second equation is estimated in levels by pooled ordinary least squares (OLS) and includes the Inverse Mill's Ratio (IMR) as an explanatory variable to correct for possible selection bias. Yet, with panel data, OLS estimates of the selected subsample are inconsistent if selection is non-random and/or if correlated individual heterogeneity is present. We thus adopt also the estimation method proposed by Wooldridge that can be used in a panel setting to take into account that there may be some unobserved time-variant factors that can affect selection and influence R&D levels through the error term. In this approach, the time-invariant effects are assumed to be linked with x^1_{it} through a linear function of k^1_i on the time averages of x^1_{it} (denoted by $x^1_i\_$bar) and an orthogonal error term a_i which exhibits no variation over time and is independent of x^1_{it} and u^1_{it}:

$$k^1_i = x^1_i\_\text{bar} + a_i$$

Hence equation (4.1) can be rewritten as:

$$D\_R\&D^*_{it} = \beta_1 x^1_{it} + \gamma_1 \bar{x}^1_i + v^1_{it} \tag{4.1a}$$

with the composite error term $v^1_{it} = u^1_{it} + a_i$ being independent of x^1_{it} and normally distributed with zero mean and variance $\sigma 2$. In this approach, to obtain estimates for the IMR, a standard probit on the selection equation (4.1a) is estimated for each t. In this approach, equation (4.2) can be rewritten as:

$$LnR\&D^*_{it}/(D\_R\&D_{it} = 1) = \beta_1 x^2_{it} + \gamma_2 \bar{x}^2_i + \zeta\lambda^2_{it} + v^2_{it} \tag{4.2a}$$

where λ_{it} is the IMR and v^2_{it} is an orthogonal residual.

Wooldridge proposes estimating equation (4.2) by including the t IMR obtained from the selection equation for each time period along with the regressors. Moreover, as the error term is allowed to be correlated with the IMR, equation (4.2a) can be consistently estimated by pooled OLS. We followed the procedure described by Wooldridge and calculated panel bootstrapped standard errors clustered by firm. This allows for obtaining standard errors corrected for first-stage probit estimates and robust to heteroskedasticity and serial correlation. The two approaches by Heckman and Wooldridge allow predicting the potential R&D for non-reporting firms.

5.2   The Knowledge Generation Function

To describe the generation process that goes from knowledge input to knowledge output, we rely on a two-equation model: the first (equation 4.3) accounts for the fact that only relatively few firms have patents, and thus analyses limited to them may be biased; the second (equation 4.4) is the knowledge generation function. Equation (4.3) is formalized as:

$$D\_PAT^*_{it} = \alpha Ln\widehat{R\&D}_{it} + \beta_3 x^3_{it} + u^3_i \tag{4.3}$$

with $D\_PAT_{it} = 1$ if $D\_PAT^*_{it} > 0$, $D\_PAT_{it} = 0$ otherwise.

D_PAT_{it} is the observed dichotomous variable taking value 1 if the firm has applied for a patent and $D\_PAT^*_{it}$ is the corresponding latent variable. Equation (4.3) is estimated using the predicted values for all firms obtained from the estimations of equations (4.1) and (4.2).[5] As a consequence, $Ln\widehat{R\&D}_{it}$ is the prediction of the R&D intensity variable from equation (4.2) and/or (4.2a), and α is the corresponding coefficient; x^3_{it} is a vector of other explanatory variables including the IMR as an explanatory variable to correct for possible selection bias.

Equation (4.3) tests the hypothesis that the probability to patent is determined by: i) the estimated levels of R&D activities; ii) the amount of external knowledge; iii) the size of the firm; and iv) the interaction term $lnR\&D\_hat *lnRegK$ that is expected to catch the multiplicative complementarity between internal and external knowledge. Note that the interaction term is expected to catch the complementarity between the current internal learning and research efforts of each agent and the current efforts of all the other co-localized firms. This specification enables us to test the hypothesis that the decision to generate new technological knowledge is contingent upon the flow of current research and learning efforts of co-localized firms. We estimate equation (4.3) using a standard probit estimator.

Finally, equation (4.4) tests the knowledge generation function (1), where the knowledge output is measured in terms of number of patents while all terms on the right-hand side enter in logarithmic form. Equation 4.4 is formalized as follows:

$$N\_PAT^*_{it} / (D\_PAT_{it} = 1) = \beta_4 x^4_{it} + u^4_{it} \tag{4.4}$$

with $N\_PAT_{it} = N\_PAT^*_{it}$ if $N\_PAT_{it} > 0$, $N\_PAT_{it} = 0$ otherwise.

The x^4_{it} are the explanatory variables and β_4 the respective coefficients.

Here, the output measure is explained by three sets of independent time-varying variables that identify the specific relevant characteristics of the size and the internal knowledge base and the amount of external knowledge respectively. The internal knowledge base here plays a central role in order to appreciate the effects of knowledge cumulability both in terms of its size, as measured by the stock of patents, and its composition, as measured by the cognitive distance (see Section 5.4 for its detailed specifications). Finally, the alternative interaction term (*lnRegK* *lnKSTOCK) is expected to catch the multiplicative complementarity between internal and external knowledge (see Section 5.3).

As the dependent variable *N_Pat*, measuring the number of a firm's patents, is a count variable, equation (4.4) is estimated using count models that prove more appropriate in dealing with non-negative integers. More precisely, equation (4.4) can be estimated by means of either a Poisson or a negative binomial model.

Since our dependent variable is over-dispersed, as shown in Table 4.3 by the fact that its variance is far larger than the mean for the sample of patenting firms, the negative binomial estimator seems to be more appropriate. Moreover, since firms included in our sample belong to all industrial sectors, they show different patenting behaviour. For this reason, equation (4.4) can be estimated by a zero-inflated regression model. Zero-inflated models attempt to account for excess zeros by means of the estimation of two equations simultaneously, one for the count model and one for the excess zeros. In other words, zero-inflated models deal with two sources of over-dispersion: a qualitative part, which explains the presence or absence of patent count; and a quantitative part, which explains the positive patent count for firms having at least one patent in a given year. Zero-inflated regression models might be a good option if there are more zeros than would be expected by either a Poisson or a negative binomial model. We thus finally use a zero-inflated negative binomial regression estimator. To account for the panel nature of our dataset, we cluster on firm identifiers to correct the standard errors within cluster similar values.

5.3   Variables' Measurement Methods

In this section we describe all variables' measurement methods.

D_R&D is a dummy taking value 1 if a firm's R&D expenditures are positive. The variable *lnR&D* is specified in relative terms as the ratio of R&D expenditures to total assets for firm i at time $t - 1$.[6]

LnIA is the logarithm of intangible assets for firm i at time $t - 1$ and is used in equations (4.1) and (4.2) to predict R&D expenditures.

In order to appreciate the effects of the internal stocks of knowledge of firms, we have included the variable *lnKSTOCK* measured in terms of the number of patents held by each firm. This is computed by applying the permanent inventory method to patent applications. We calculate it as the cumulated stock of patent applications using a rate of obsolescence of 15 per cent per annum. The choice of the rate of obsolescence raises some basic issues as to the most appropriate value. There are indeed a number of studies moving from Schankerman and Pakes (1986) that attempted to estimate the patent depreciation rate. However, for the scope of this chapter we follow the established body of literature based on Hall et al. (2005) that applies to patent applications the same depreciation rate as the one applied to R&D expenditures (see for example McGahan and Silverman, 2006; Coad and Rao, 2006; Nesta, 2008; Laitner and Stolyarov, 2013; Rahko, 2014).

$$KSTOCK_{it} = \dot{h}_{it} + (1-\sigma)KSTOCK_{it-1} \qquad (4.5)$$

where $\dot{h}_{it}$ is the flow of patent applications and σ is the rate of obsolescence. We finally compute the logarithm of the patent stock.

The variable $lnCD$ accounts for the composition of the stock of knowledge internal to each firm, and qualifies its knowledge base in terms of cognitive distance ($lnCD$). Following the recombinant knowledge approach, this index expresses knowledge dissimilarities among different types of knowledge (see Section 5.4 for the specifications and measure of this variable).

The third set of variables accounts for the contribution of the knowledge that is external to each firm at time t but made accessible by proximity. Here $lnRegK$ measures the patenting activities of firms localized within the same region, and as such can produce positive pecuniary knowledge externalities mainly based upon the mobility of skilled personnel and more generally of the array of knowledge interactions that make the access and use of external knowledge cheaper.

We also include a dedicated variable to account for the contribution of the knowledge that is external to each firm but made accessible by inter-regional knowledge flows. Here $lnExtRegK$ measures the patenting activities of firms localized outside their region and aims at capturing the role of the sources of external knowledge that are far away from firm i. The variable has been computed as the number of patents (in Log) in the NUTS2 regions of the EU-24 member states, weighted using distance from firm i's region at time $t - 1$.

Moreover, to catch the multiplicative cumulability between internal and external knowledge in equation (4.3), we develop the interaction term $lnR\&D\_hat *lnRegK$. Specifically, in this interaction term the levels of estimated R&D activities at the firm level multiply the levels of external knowledge in terms of patenting activities.

In addition, to catch the multiplicative cumulability between internal

Table 4.2   Variables' measurement methods

	Variable	Measurement method
R&D dummy	$D\_R\&D$	Dummy = 0 if firm's R&D expenditures are positive
R&D intensity	$lnR\&D$	Log (R&D / total assets) for firm i at time $t-1$
Intangible assets	$lnIA$	Log intangible assets for firm i at time $t-1$
External knowledge within the firm's region	$lnRegK$	Log of no. of patents in the same region (NUTS2) of firm i at time $t-1$
External knowledge outside the firm's region	$lnExtRegK$	Log of no. of patents in regions (NUTS2) belonging to EU-24 member states other than that of firm i at time $t-1$, weighted using distance
Patent dummy	D_Pat	Dummy = 1 if firm has applied for patents
Number of patents	N_Pat	No. of patents for firm i at time t
Knowledge stock	$lnKSTOCK$	Log of patent stocks (PIM) for firm i at time $t-1$
Cognitive distance	$lnCD$	Log of cognitive distance of firm i at time $t-1$
Firm's size	$lnSize$	Log of deflated sales for firm i at time $t-1$

and external knowledge in equation (4.4), we develop the interaction term $lnRegK *lnKSTOCK$. Here, the stock of knowledge at the firm level multiplies the levels of external knowledge. This specification enables us to test the hypothesis that the amount of knowledge output that a firm is able to generate is contingent upon the multiplicative relationship between the internal stock of knowledge and the levels of current efforts of all the other co-localized firms.

We finally included a set of control variables. The variable $lnSize$ controls for the firm's size and is measured as the Log of its sales (deflated using industry deflators). Moreover, we included both sectoral and time dummies in order to control for industrial and time effects. For each variable the measurement method is defined in Table 4.2, while descriptive statistics are reported in Table 4.3. The correlation matrix can be found in Table 4.4.

5.4   The Cognitive Distance Index

The implementation of the cognitive distance index, proxying for similarity, rests on the recombinant knowledge approach. In order to provide

Table 4.3 Descriptive statistics

Variable	Total			R&D reporting			Patenting		
	Obs	Mean	Std Dev.	Obs	Mean	Std Dev.	Obs	Mean	Std Dev.
lnR&D	6305	-3.602	1.658	6305	-3.602	1.658	2503	-3.369	1.424
lnIA	16,998	8.639	3.124	6305	9.518	3.128	3511	9.488	3.138
lnRegK	16,998	6.479	1.457	6305	6.429	1.393	3511	6.608	1.380
lnExtRegK	16,998	5.462	0.254	6305	5.463	0.259	3511	5.454	0.275
N_PAT	16,998	1.322	1.400	6305	3.332	22.66	3511	6.400	30.287
lnKSTOCK	3083	1.460	1.813	2254	1.798	1.880	3083	1.460	1.813
lnCD	886	-4.255	1.212	793	-4.272	1.199	885	-4.258	1.209
lnSize	16,998	11.446	2.535	6305	12.069	2.833	3511	12.409	2.913

Table 4.4 Correlation matrix

	lnR&D	N_PAT	lnIA	lnRegK	lnExtRegK	lnKSTOCK	lnCD	lnSize
lnR&D	1							
N_PAT	0.0448	1						
lnIA	-0.353	0.3009	1					
lnRegK	-0.0373	0.0817	0.3108	1				
lnExtRegK	-0.0277	-0.0303	0.0521	-0.0522	1			
lnKSTOCK	-0.0922	0.5772	0.5991	0.2185	0.2109	1		
lnCD	0.1739	-0.3032	-0.3084	-0.1985	0.1384	-0.4157	1	
lnSize	-0.3857	0.3545	0.7976	0.2281	-0.0049	0.6677	-0.3854	1

an operational translation of such a concept one needs to identify both a proxy for the bits of knowledge and a proxy for the elements that make up their structure. For example, one could take scientific publications as a proxy for knowledge, and look either at keywords or at scientific classification (like the JEL code for economists) as a proxy for the constituting elements of the knowledge structure. Alternatively, one may consider patents as a proxy for knowledge, and then look at technological classes to which patents are assigned as the constituting elements of its structure, that is, the nodes of the network representation of recombinant knowledge. Here we will follow this latter avenue. Each technological class *l* is linked to another class *j* when the same patent is assigned to both of them.[7] The higher the number of patents jointly assigned to classes *l* and *j*, the stronger is this link. Since technological classes attributed to patents are reported in the patent document, we will refer to the link between *l* and *j* as the co-occurrence of both of them within the same patent document.[8]

On this basis we calculated the cognitive distance (CD), which expresses the average degree of dissimilarity among different types of knowledge (Nooteboom, 2000). A useful index of distance can be derived from *technological proximity* proposed by Jaffe (1986, 1989), who investigated the proximity of firms' technological portfolios. Breschi et al. (2003) adapted this index to measure the proximity or relatedness between two technologies.

We define $P_{lk} = 1$ if the patent *k* is assigned the technology *l* [*l* = 1, . . ., n], and 0 otherwise. The total number of patents assigned to technology *l* is $O_l = \sum_k P_{lk}$. Similarly, the total number of patents assigned to technology *j* is $O_j = \sum_k P_{jk}$. We can, thus, indicate the number of patents that are classified in both technological fields *l* and *j* as $V_{lj} = \sum_k P_{lk} P_{jk}$. By applying this count of joint occurrences to all possible pairs of classification codes, we obtain a square symmetrical matrix of co-occurrences whose generic cell V_{lj} reports the number of patent documents classified in both technological fields *l* and *j*.

Technological proximity is proxied by the cosine index, which is calculated for a pair of technologies *l* and *j* as the angular separation or un-centred correlation of the vectors V_{lm} and V_{jm}. The similarity of technologies *l* and *j* can then be defined as:

$$S_{1j} = \frac{\sum_{m=1}^{n} V_{1m} V_{jm}}{\sqrt{\sum_{m=1}^{n} V_{1m}^2} \sqrt{\sum_{m=1}^{n} V_{jm}^2}} \tag{4.6}$$

The idea behind the calculation of this index is that two technologies, j and l, are similar to the extent that they co-occur with a third technology, *m*. Such a measure is symmetric with respect to the direction linking technological classes, and it does not depend on the absolute size of the technological field. The cosine index provides a measure of the similarity between two technological fields in terms of their mutual relationships with all the other fields. S_{lj} is the greater the more two technologies *l* and *j* co-occur with the same technologies. It is equal to one for pairs of technological fields with identical distribution of co-occurrences with all the other technological fields, while it goes to zero if vectors V_{lm} and V_{jm} are orthogonal (Breschi et al., 2003). Similarity between technological classes is thus calculated on the basis of their relative position in the technology space. The closer technologies are in the technology space, the higher is S_{lj} and the lower their cognitive distance (Breschi et al., 2003; Engelsman and van Raan, 1994; Jaffe, 1986).

The cognitive distance between *j* and *l* can therefore be measured as the complement of their index of technological proximity:

$$d_{lj} = 1 - S_{lj} \qquad (4.7)$$

Having calculated the index for all possible pairs, it needs to be aggregated at the firm level to obtain a synthetic index of technological distance. This is done in two steps. First we compute the weighted average distance (WAD) of technology *l*, that is, the average distance of *l* from all other technologies:

$$WAD_{lt} = \frac{\sum_{j \neq 1} d_{lj} P_{jt}}{\sum_{j \neq 1} P_{jt}} \qquad (4.8)$$

where P_j is the number of patents in which the technology *j* is observed. The average cognitive distance at time *t* is obtained as follows:

$$CD_t = \sum_l WAD_{lt} \times \frac{P_{lt}}{\sum_l P_{lt}} \qquad (4.9)$$

In our model the variable is included in logarithm (*lnCD*).

6. RESULTS

Table 4.5 shows the results for the R&D equations estimated using the Heckman and the Wooldridge procedures. More precisely, the estimates of

Table 4.5   Estimation results for the R&D equations

Equation (4.1) Dep. Var.	(4.1) D_R&D	(4.1a) D_R&D
lnINTASS	0.1044***	See Antonelli and
		Colombelli (2015a)
	(0.0054)	
lnSize	0.0312***	
	(0.0066)	
Time dummies	Yes	
Sectoral dummies	Yes	
N	16,998	

Equation (4.2) Dep. Var.	(4.2) Pooled OLS *lnRED*	(4.2a) Wooldridge *lnRED*
lnINTASS	0.3817***	−0.0852***
	(0.0902)	(−4.27)
lnINTASS_bar		0.139***
		(4.24)
Time dummies	Yes	Yes
Sectoral dummies	Yes	Yes
Constant	−15.0707***	−7.892***
	(2.0508)	(−10.40)
IMR	6.3322***	2.087***
lambda	(1.1110)	(4.93)
N	6305	6305

Notes:
Standard errors in parentheses.
* $p < 0.10$, ** $p < 0.05$, *** $p < 0.01$ (here and in Tables 4.6–4.7).

equations (4.1) and (4.2) are reported in columns (4.1) and (4.2), while the estimates of equation (4.2a) are reported in column (4.2a).[9] The two sets of equations allow predicting the potential R&D for non-reporting firms.

Table 4.6 shows the estimation results for equation (4.3) and the extended one that includes the interaction variable. Columns (4.3) and (4.3') show results if the levels of R&D are estimated using the Heckman procedure, and columns (4.3a) and (4.3a') if the levels of R&D are estimated using the

Table 4.6   Estimation results for Equation 4.3

Equation (4.3) Dep. Var.	(4.3) D_PAT	(4.3') D_PAT	(4.3a) D_PAT	(4.3a') D_PAT
lnRED_hat	0.119*** (29.93)		0.153*** (11.11)	
lnRegK	0.0984*** (10.57)		0.0851*** (9.50)	
lnRED_hat * *lnRegK*		0.0122*** (21.89)		0.00296* (1.87)
lnExtRegK	0.223*** (3.46)	0.140** (2.25)	0.0411 (0.66)	0.0486 (0.79)
lnSize	0.0668*** (12.48)	0.0949*** (18.30)	0.124*** (22.50)	0.115*** (21.11)
Time dummies	Yes	Yes	Yes	Yes
Sectoral dummies	Yes	Yes	Yes	Yes
Constant	−2.461*** (−6.52)	−2.040*** (−5.68)	−1.6524*** (0.4143)	−2.379*** (−6.65)
N	16,998	16,998	16,998	16,998

Wooldridge procedure. Our results confirm the hypothesis that the probability to patent is determined by: i) the estimated levels of R&D activities; ii) the amount of external knowledge; iii) the size of the firm; and iv) the complementarity between internal and external knowledge. First, R&D activities contribute significantly to the generation of new knowledge, as confirmed by the positive and significant coefficient of *lnRED_hat*. Second, the positive and significant role of external knowledge is confirmed by the results of the *lnRegK* variable. Moreover, in order to check further the role of external knowledge, we also included in our model a variable measuring the patenting activities of firms localized outside their region (*lnExtRegK*). *lnExtRegK* turns out to be significantly related to the probability of generating new technological knowledge only in two out of four regressions. Finally, the interaction term proved to be positively and significantly related to a firm's probability to patent (columns 4.4' and 4.4a').

Table 4.7 shows the results of the zero-inflated negative binomial regression estimations for equation (4.4) and the extended one that includes the interaction variable. Columns (4.4) and (4.4') show results if the levels of R&D are estimated using the Heckman procedure, while columns (4.4a)

Table 4.7   Estimation results for Equation 4.4

Equation (4.4) Dep. Var.	(4.4) N_PAT	(4.4') N_PAT	(4.4a) N_PAT	(4.4a') N_PAT
LnCD * lnKSTOCK	−0.1303*** (0.0088)	−0.0784*** (0.0099)	−0.1289*** (0.0082)	−0.0808*** (0.0097)
lnRegK * lnKSTOCK		0.0503*** (0.0086)		0.0521*** (0.0089)
lnSize	0.0739** (0.0328)	0.0733** (0.0349)	0.0002 (0.0421)	−0.5929*** (0.2102)
IMR lambda	−0.5543* (0.2980)	0.1427 (0.2892)	−1.3239*** (0.4049)	−9.0380*** (2.6659)
Time dummies	Yes	Yes	Yes	Yes
Sectoral dummies	Yes	Yes	Yes	Yes
Constant	−3.3782*** (0.6352)	−4.3537*** (0.6185)	−1.3670 (0.9805)	15.4298*** (5.9484)
Inflate macro-sector	1.8629 (1.3856)	0.0803 (0.3026)	0.0197 (0.2839)	1.7295 (1.4122)
Constant	−0.5183*** (0.1220)	−0.7274*** (0.1336)	−3.6444*** (1.2266)	−14.1992 (8.9442)
lnalpha				
Constant	−0.5172*** (0.1269)	−0.6558*** (0.1341)	−0.6150*** (0.1319)	−0.7736*** (0.1319)
N	883	883	883	883

and (4.4a') if the levels of R&D are estimated using the Wooldridge procedure. The negative and significant results of the knowledge base – specified as the multiplicative relation between the stock of patents and their cognitive distance (*LnCD * lnKSTOCK*) – suggests, first, that stocks of patents with high levels of similarity exert positive effects on the generation of new knowledge. This means that when firms focus their search activity in a region of the knowledge space that is close to their accumulated competences, they are more likely to achieve the successful generation of new knowledge. Second, the interaction term (*lnRegK * lnKSTOCK*), designed to catch the multiplicative interaction between internal and external knowledge, exerts a positive and significant impact on the generation of new knowledge.

These results fully confirm our hypothesis that the generation of new technological knowledge is affected by the complementarity between internal and external knowledge, and by the composition and size of the stock of internal knowledge.

7.   CONCLUSIONS

The knowledge generation function is an important tool that enables us to open the black box of the knowledge generation process. The Arrovian analysis of the characteristics of knowledge as an economic good has important implications to understand the specificities of the knowledge generation process. The key attributes of indivisibility and limited appropriability identified – especially when the twin character of knowledge that is at the same time the output of the generation process and an input in the following generation process – is fully grasped, together with its implications.

Analysis of knowledge indivisibility has made it possible to identify the two key dimensions of internal and external knowledge cumulability. The efficient and effective generation of new knowledge at time t is possible only standing on the shoulders of the technological knowledge that has been generated until that time both by the very same firm and by the other firms in geographical, technological and industrial proximity. Since no agent can command all the technological knowledge that has been generated at each point in time and the limited appropriability of technological knowledge engenders the possibility to access, absorb and use the knowledge that has been generated by third parties, the knowledge external to each firm plays a key role in the generation of new technological knowledge.

The early specifications of the knowledge generation function did not pay attention to the characteristics of knowledge as an economic good and could not take advantage of their important implications to understand the dynamic process that makes the generation of new technological knowledge possible. The results of our empirical analysis confirm that the output of knowledge is generated not only by means of R&D expenditures: knowledge external to each firm, the stock of existing knowledge within the firm and the composition of the knowledge internal and external to each agent enter the knowledge generation function as essential inputs. Our results suggest the strong and positive role of the stock of knowledge possessed by each firm, and the key role of proximity. External knowledge plays a role as an input when it is embedded in firms that are co-localized in regional proximity. Proximity also matters with respect to the components of the knowledge base: the higher the similarity of the knowledge base, the larger the output in terms of new knowledge.

These results have important implications on three counts. First, they confirm that the generation of technological knowledge is the result of a systemic process that is jointly influenced by the efforts of each agent, by the size and the composition of the stock of knowledge cumulated in the past, and by the amount of external knowledge that can be accessed within the system in which each firm is embedded. These results help put the excess attention paid to R&D expenses in the appropriate context, grasping the role of the other essential factors in the knowledge generation process.

The specific endowment cumulated in the past through time and represented by the size and the composition of the stock of technological knowledge exerts long-term effects on the actual ability of firms to generate new technological knowledge. These effects can be altered and affected by the current effects of R&D expenditures that each firm is able to fund and perform at each point in time and by the quality of the pools of external knowledge that each firm can access.

Technological knowledge can no longer be regarded as a homogeneous bundle. Its composition in terms of specific components must be appreciated and taken into consideration both from a policy and a strategy viewpoint. The identification of the actual levels of complementarity of the types of knowledge available in a region and in the stock of other firms must be closely assessed when public interventions to support the introduction of innovations and strategic action by firms are at stake.

Appreciation of the key role played by external knowledge enables us to fully understand the systemic conditions that shape and make the generation of new technological knowledge possible. The generation of new technological knowledge is influenced by individual characteristics of each firm, such as the past accumulation of knowledge and the current commitment of resources to research, but requires access to the complementary knowledge that is external to each firm because it has been generated and is – partly – possessed by other firms in geographical, industrial and technological proximity. The generation of technological knowledge cannot be regarded as the result of stand-alone activity, but rather as the product of a collective process. This leads to the identification of innovation as an emergent property of a system. The characteristics of the system are crucial to assess the amount and the characteristics of the knowledge being generated.

These results are important for their implications for both public policy and corporate strategy. Appreciation of the strong complementarity of pecuniary knowledge externalities stemming from external knowledge and current inputs, together with the internal stocks of knowledge, draws attention to the fact that the actual amount of knowledge that a firm is able to generate with given levels of R&D is heavily influenced by the size and density of the local pools of knowledge as shaped by the knowledge possessed by other firms and by the networks in which firms are able to participate.

NOTES

1. The knowledge generation function has very often been called the knowledge production function (KPF). Former use of this latter terminology can be found in Pakes and Griliches (1984). Jaffe (1986) instead calls it patent equation. However, there is also much literature that used to label KPF the technology production function, engendering some confusion. We therefore use the label 'knowledge generation function' so as to rule out any possible ambiguity.
2. See Griliches (1979: 95, note 3): 'An alternative approach would complicate this model further by adding an annual knowledge production function of the form $dK = H(R, K)$ and defining K accordingly' (where K stands for knowledge and R for R&D activities). Griliches' subsequent contributions (1990 and 1992) fully confirm that his main interest was assessment of the statistical merits of patents as an indicator of R&D rather than exploration of the relationship between patents and R&D expenditures as a knowledge generation function.
3. See Jaffe (1986: 988): 'I begin by assuming that the new knowledge produced by the firm in any period is related to its R&D in that period according to a modified Cobb–Douglas technology . . . where k, is the new knowledge generated by firm i, r, is its R&D spending, and s, is the potential "spillover pool" whose construction is described above.'
4. Implementation of the IPER database was financed by the Collegio Carlo Alberto, under the IPER project.
5. This approach reflects the assumption that all firms carry out innovative activities, although some of them do not report any innovative investment.
6. As in our model, we do include the variable in logarithmic form; we obtain negative values in the descriptive statistics of *lnR&D* and *lnCD*.
7. In the calculations four-digit technological classes have been used.
8. It must be stressed that to compensate for intrinsic volatility of patenting behaviour, each patent application lasts five years in order to reduce the noise induced by changes in technological strategy.
9. The *t* estimates of equation (4.1a) are not given here but are reported in Antonelli and Colombelli (2015a).

5. The role of external knowledge in the introduction of product and process innovations

Cristiano Antonelli and Claudio Fassio

1. INTRODUCTION

This chapter contributes to the literature on knowledge generation with a novel approach that stresses the heterogeneity of the sources of external knowledge and their differentiated effects on types of innovation. The crucial role of external knowledge in the generation of new knowledge as a necessary, complementary input in the generation of new technological knowledge is well known. External knowledge however cannot be regarded as a homogeneous bundle. Its origins, properties and acquisition mechanisms are major factors that determine to what extent firms can actually benefit from access to the external inputs of knowledge. Indeed, not all types of external knowledge convey the same innovative outputs, and firms should update their knowledge-sourcing strategies accordingly.

This chapter focuses on the inbound side of knowledge sourcing and open innovation, and makes two contributions. First we show that different sources of knowledge impact different types of innovation. More specifically, we distinguish among: a) horizontal sources of knowledge stemming from competitors in the same industry; b) transaction-based upstream vertical knowledge embodied in acquisitions of machinery; c) upstream vertical knowledge flows proceeding from interactions with suppliers; d) vertical downstream knowledge stemming from the interaction with customers. We show how these different types of external knowledge impact differently on product and process innovation. While horizontal and vertical downstream sources have a stronger effect on product innovation, upstream vertical knowledge (whether stemming from transactions or interactions with suppliers) mainly favours process innovation. We also show how these findings are surprisingly similar for different types of firms: even when we consider separately small, medium and large firms, as well as high-tech and low-tech ones, our results remain largely unchanged.

The implications of these findings are the second contribution of this chapter. Awareness of the differentiated effects of the various sources of external knowledge on firms' innovative performance should push managers to act strategically in their inbound exploratory practices. Managers should be able to identify which sources of external knowledge are available in the system in which firms are embedded and direct firms' innovative strategies accordingly in order to benefit as much as possible from those specific sources of knowledge.

Accounting for the heterogeneity of external knowledge seems an important research topic on which major progress can be done, through exploration of the diversity of technological knowledge according to its sources and its effects. The rest of the chapter is organized as follows. Section 2 reviews the related literature and introduces the hypotheses, while Section 3 presents the methodology for the empirical investigation, the data sources and the results of the econometric analyses. The conclusions summarize the results and outline the main implications of the analysis.

2.   THEORETICAL FRAMEWORK AND HYPOTHESES

Analysis of the knowledge generation process has been at centre stage of the economics of innovation for quite some time. Building upon the important acquisitions of the analysis of the characteristics of knowledge as an economic good, it has been possible to explore the economic activities that lead to its generation and eventual application through the introduction of technological innovations (Griliches, 1979; Pakes and Griliches, 1984; Jaffe, 1986). The central role of external knowledge has been identified and appreciated as an indispensable complementary input, together with R&D activities, in the recombinant generation of new technological knowledge: no firm can generate the technological knowledge that is necessary to innovate, alone, in isolation (Schumpeter, 1947; Weitzman, 1996; Antonelli, 2008).

Arora and Gambardella (1990) and Cassiman and Veugelers (2006) confirm the complementarity of external knowledge and highlight the limited possibility to substitute it with internal R&D. Complementarity also implies a system-level perspective, since agents with little R&D cannot benefit from the opportunities offered by regions rich in technological spillovers, while major R&D efforts may fail when the level of knowledge externalities in the system is not appropriate (Griffith et al., 2003;

Ruckman, 2008; Johansson and Lööf, 2015). Antonelli and Colombelli (2015a) articulate and test the hypothesis that external and internal knowledge are complementary as each of them is indispensable and cannot fall to zero levels without annulling output.

The technology management literature has provided substantial contributions to understanding the role of external knowledge in knowledge generation implementing the open innovation research agenda (Enkel et al., 2009). This literature has stressed the role of the inbound open innovation strategies that consist in the acquisition of external knowledge inputs and their systematic integration with internal knowledge in knowledge generation (Chesbrough, 2003; Chesbrough and Crowther, 2006; Bianchi et al., 2010; Dahlander and Gann, 2010; Mortara and Minshall, 2011).

Identification of the variety of activities that make access to external knowledge possible and appreciation of their differentiated effects on the rate of introduction of innovations have been among the main results of this line of enquiry. However the results of these studies have also highlighted the potential pitfalls of inbound open innovation strategies. Howells et al. (2008) and Grimpe and Kaiser (2010) study in detail the effects of one such activity to access external knowledge, that is, R&D outsourcing. They identify relevant drawbacks as they note that it is necessary to take into account the relevant cost of its integration within the internal knowledge generation process with potential negative consequences in terms of dilution of firm-specific resources. Grimpe and Kaiser identify a maximum level of R&D outsourcing beyond which over-outsourcing has negative effects on the rate of introduction of innovations. Ebersberger and Herstad (2011) explored in detail the effects of three modes of acquisition of external knowledge – search, collaboration and external R&D – on the introduction of product innovations. Search and collaboration affect innovation performance positively, and are complementary to each other. R&D outsourcing instead seems to exert a negative impact that is partly reduced if it is associated with search and, conversely, reinforced when it takes place with collaboration (Lin et al., 2012).

Theyel explores the different components of the value chain of firms to assess the stage and the mode of access and use of external knowledge. According to her results more than 50 per cent of firms rely on external knowledge during technology and product development and commercialization, while only 30 per cent of them engage in joint manufacturing. Theyel suggests that open innovation practices do not always lead to improvements in innovation performance; in some circumstances firms reach a 'tipping point' where 'the usage of additional external sources after a certain amount may become a burden rather than an advantage, resulting in negative returns' (Theyel, 2013: 261).

These results cast some doubt on the effective role of external knowledge on the rate of introduction of technological innovations and recall the crucial need to balance internal sources with external ones (March, 1991). As Faems et al. (2010) note, inbound open innovation strategies are highly context dependent. Their implementation and actual success is heavily affected by the characteristics of the system in which firms are embedded (see also Lee and Wong, 2011).

Hypotheses

These last results suggest that a useful improvement on the existing literature consists in an analysis that does not merely focus on the quantitative contribution of external knowledge to innovative outcomes, but rather investigates the variety of sources of external knowledge and their differentiated impact on different types of innovation outputs. Here we will consider the effect of a number of different sources of external knowledge on innovative outcomes. As Di Stefano et al. (2012) remark, the old dilemma between technology-push and demand-pull inducements on innovation suggest that relevant knowledge inputs for the development of innovations might stem both upstream from suppliers and downstream from customers. We analyse the role of external knowledge that flows vertically across industries, both upstream from suppliers and downstream from clients, as well as external knowledge inputs stemming horizontally from competitors in the same industry. The sources of external knowledge are likely to influence not only the amount of new technology being introduced, but its type: specifically, we focus on their effect on process and product innovations. Let us articulate our hypotheses in detail.

Cappelli et al. (2014) show that horizontal knowledge spilling from competitors active in the same product market seems better able to contribute to the generation of new technological knowledge directed towards the introduction of product innovations. Here external knowledge consists in the reciprocal borrowing and use of technological knowledge produced by each competitor able to imitate and implement the innovations introduced by the other firms engaged in the same innovation race (see also Laursen and Salter, 2006; Quatraro, 2009; O'Regan and Kling, 2011). On the basis of this evidence we put forward our first hypothesis:

H1: Horizontal flows of external knowledge that spill among competitors in the same product markets would favour the generation of new technological knowledge with a bias in favour of product innovations.

Technological knowledge also flows vertically among firms within value chains, across industries. Transactions between suppliers and customers of capital goods and other intermediary inputs are a primary source of embodied technological knowledge that affects the innovation process of downstream users. Firms that acquire advanced capital goods that embody new technological knowledge take advantage of learning by using, and introduce new technologies in their production process (Parisi et al., 2006). In a similar way, user–producer interactions between potential innovators and their suppliers activate upstream knowledge flows that empower the access to tacit knowledge and competence on the characteristics of the production process: again this favours, mainly, the introduction of process innovations (Fassio, 2015). Accordingly, our second hypothesis is the following:

H2: External knowledge that flows vertically upstream within user–producer relations should mainly favour the generation of technological knowledge for process innovations.

Bartelsman et al. (1994) distinguish between upstream and downstream flows of technological knowledge. User–producer interactions between potential innovators and their customers activate downstream knowledge flows that empower the access to tacit knowledge and competence on the characteristics of the products so as to favour – mainly – the introduction of product innovations. As also shown by the literature on user-driven innovation (von Hippel, 1998), it is likely that external knowledge stemming from customers of technological products will favour the introduction of product innovations thanks to the competent knowledge provided by users. Our third hypothesis is hence the following:

H3: External knowledge that flows vertically downstream within user–producer interactions should mainly favour the generation of technological knowledge for product innovations.

Table 5.1 and Figure 5.1 provide a graphical representation of our hypotheses. The matching of types of innovation and sources of external technological knowledge is crucial to understanding properly the actual role of external knowledge and to design both policy instruments and strategic action.

Table 5.1   Types of external knowledge and innovation

	Horizontal external knowledge	Vertical external knowledge
Product innovation	When competitors in the same product markets are the main source of external knowledge the innovation processes of firms rely on creative imitation that favours mainly the introduction of new products	Downstream user–producer interactions with clients are an important source of tacit knowledge and competence about the characteristics of the products, and hence a major source of technological knowledge that enables the introduction of product innovations
Process innovation		Process innovations are favoured by: i) high-quality upstream use–producer interactions with suppliers; ii) relevant transactions of advanced machinery and intermediary inputs

3.   THE EMPIRICAL ANALYSIS

3.1   Data

In order to test empirically the relevance of the different sources of external knowledge to product and process innovation we use Eurostat's harmonized Community Innovation Survey (CIS) 4 data, which refer to the period 2002–2004.[1] The great advantage of this database is the possibility to treat company data from the same wave of the CIS together for different countries. We hence pooled in a unique database all the firms that answered the questionnaire from all the manufacturing sectors. We limit our analysis to the six larger countries included in the database: Belgium, Czech Republic, Germany, Italy, Norway and Spain.

Dependent variables
In the CIS each firm is asked whether it has introduced at least one product or process innovation in the time span covered by the survey. The dependent variables are: a dummy variable for the introduction of process innovation; another dummy for the introduction of product innovation; and a third dummy for the introduction of at least one of the two types of innovation.

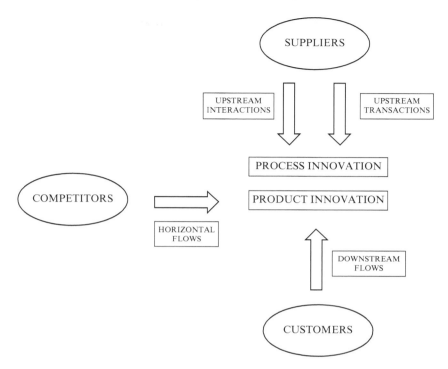

Figure 5.1   Differentiated effects of sources of innovation

Independent variables

Horizontal flows of external knowledge   The horizontal flows of external knowledge measure the external knowledge coming from the industrial environment in which the firm is embedded. In CIS4 the firms surveyed are asked to report the total amount of internal and external R&D in the time span considered. On the basis of this data we summed up all the firms' expenditures in R&D for each national sector and obtained the overall amount spent in that sector, hence an industry-specific variable. More specifically:

$$HORIZONTAL_j = \sum_i RD_{ij}$$

where j = sector of firm i.

In order to take into account potential size effects due to cross-country differences we also computed the average level of R&D for each sector by dividing the total sum of R&D by the number of firms in each sector.

Downstream vertical flows of external knowledge   Downstream vertical flows are measured by the interaction between firms and their customers. The flows of external knowledge proceeding from interactions with users and customers are proxied by a dummy variable that is equal to 1 if a firm declares that customers have been an important source of information for the development of innovations.

Transaction-based upstream vertical flows   The flows of external knowledge that firms are able to access through transactions with their suppliers are measured by the share of expenditures in machinery acquisition over the total sales by each firm. The purchase of machinery is the primary means of acquisition of external knowledge embodied in capital goods.

Interaction-based upstream vertical flows   The flows of knowledge stemming from interactions with upstream suppliers are proxied by a dummy variable that is equal to 1 if a firm declares that suppliers are an important source of information in the innovative process.

Other controls
We also control for the use of internal sources of knowledge by firms, and use a standard measure of internal R&D. Moreover, we control for company size, distinguishing between small, medium and large firms. We also check the sectoral affiliation of each firm, using the two-digit sector ISIC rev.3. classification. Our database is composed of all the manufacturing firms who answered the CIS4 survey in the six countries considered. After some necessary cleaning procedures to eliminate outliers our database is made of 23,247 observations, which include both innovating and non-innovating firms.[2] Table 5.2 provides some descriptive statistics of the firms in the database.

3.2   Methodology

To test our hypotheses we start from a generic knowledge generation function in which (IK) and (EK) are vectors of indispensable and complementary inputs in the generation of new knowledge (KN):

$$KN = f \ (IK \ EK) \tag{5.1}$$

We then elaborate the basic knowledge generation function with the following specifications: i) internal knowledge is proxied by the firm's own investments in in-house R&D; ii) external knowledge is distinguished in its four heterogeneous components – downstream stemming from interac-

Table 5.2   Descriptive statistics

Variables	Mean (1)	St. dev. (2)
Firm-level variables		
upstream-interaction	0.123	0.328
upstream-transaction (logs)	−9.348	3.381
downstream	0.157	0.364
horizontal*O_i (sum of sectoral R&D in logs)	3.287	3.876
horizontal*O_i (average sectoral R&D in logs)	2.260	2.775
O_i (predicted values)	0.176	0.196
R&D intensity (logs)	−1.618	2.302
Turnover in 2004 (logs)	15.59	1.793
size (employees): < 50	0.552	0.497
size (employees): 50–249	0.307	0.461
size (employees): > 250	0.141	0.348
Patenting activity	0.146	0.353
Local funding	0.121	0.326
National funding	0.124	0.330
European funding	0.055	0.308
Cooperation	0.168	0.374
Industry-level variables		
Sum of sectoral R&D (logs)	17.97	1.693
Average sectoral R&D (logs)	12.00	1.632

Notes: The sample includes 23,247 firms. All financial variables are in Euros.

tions with customers (*DOWNSTR*), vertical upstream stemming from interactions with suppliers (UPSTR-*inter*), vertical upstream mediated by transactions (*UPSTR*-trans) and horizontal (*HORIZONTAL*); the output *KN* is distinguished in terms of product innovations, process innovations and innovations that affect both products and processes. Equation (5.1) is consequently re-specified as it follows:

$$KN_i^s = R\&D_i^\alpha\, DOWNSTR_i^{\beta_1}\, UPSTR - trans_i^{\beta_2} UPSTR - inter_i^{\beta_3}$$
$$HORIZONTAL_j^\gamma \tag{5.2}$$

where s = *PRODUCT INNO, PROCESS INNO, INNOVATION*.

KN denotes the knowledge generated by firm i: we distinguish among *PRODUCT INNO* and *PROCESS INNO*, which indicate the introduction of product or process innovations, while *INNOVATION* stands for the introduction of either or both of them.

Following Griffith et al. (2006b) we assume that the appropriation of the horizontal flows of external knowledge is filtered by the degree to which a firm is open to external knowledge; that is, the influence of the horizontal external stock of knowledge (*HORIZONTAL*) on firm *i* is a linear function of a firm-specific degree of openness to external knowledge, which we indicate as *O*:

$$\gamma_i = \pi O_i \qquad (5.3)$$

We proxy the openness O_i of each firm with the probability that it will invest in external R&D. We introduce a probit equation that identifies the determinants of this decision, where the dependent variable is a dummy equal to 1 if a firm invested in extramural R&D, and equal to zero otherwise. By estimating the probit model we can obtain the predicted probabilities of investing in external R&D. This new variable, which will be bounded between zero and one, will be used as a proxy for the openness variable O_i, that is, the firm-specific propensity to engage in external R&D.

We then take logs in equation (5.2), where lower case letters indicate logs:

$$kn_i^s = \alpha \, rd \, \text{int}_i + \beta_1 \, \text{downstr} + \beta_2 \, \text{upstr-trans} + \beta_3 \, \text{upstr-inter}_i + \gamma \, horizontal_j \qquad (5.4)$$

Then we simply substitute (5.3) in (5.4) and obtain the following equation to estimate:

$$kn_i^s = \alpha \, rd \, \text{int}_i + \beta_1 \, \text{downstr} + \beta_2 \, upstr\text{-}trans + \beta_3 \, upstr\text{-}inter_i +$$
$$\pi \, (horizontal_j^* O_i) + \delta x_i + u_i \qquad (5.5)$$

Equation (5.5) allows us to estimate the direct effect of the investments in internal R&D on the generation of knowledge. Moreover, the model measures the impact of the heterogeneous types of external knowledge on the innovative capacity of firms. The *x* variables are standard controls for firm size and for sector and country effects, while *u* is an idiosyncratic error term. We estimate equation (5.5) with a probit model using three proxies of knowledge creation: product innovation, process innovation or both of them.

3.3   The Results

Before estimating equation (5.5) we first build our measure of openness (O_i) for all the firms in our sample. Column (1) of Table 5.3 presents the

Table 5.3   Probit and Tobit estimation of R&D

Dependent variable	Engagement in external R&D	Engagement in R&D	Intensity of R&D
	Probit	Tobit	
	(1)	(2)	(3)
Turnover in 2004 (in logs)	0.033***	0.052***	−0.282***
	(0.002)	(0.003)	(0.018)
Belonging to a group	0.034***	0.008	0.102***
	(0.006)	(0.007)	(0.037)
International markets	0.036***	0.121***	0.214***
	(0.005)	(0.006)	(0.051)
Patenting activity	0.099***	0.276***	0.630***
	(0.009)	(0.010)	(0.042)
Size			
50–249	−0.002	0.042***	−0.115**
	(0.006)	(0.008)	(0.047)
> 250	−0.003	0.061***	0.162**
	(0.010)	(0.016)	(0.076)
Local funding	–	–	0.204***
			(0.035)
National funding	–	–	0.440***
			(0.035)
European funding	–	–	0.180***
			(0.031)
Cooperation	–	–	0.207***
			(0.034)
Constant	–	–	−0.792***
			(0.303)
Observations	23,247	23,247	23,247
Wald test of indep. eqns (rho = 0)	–	142.1	142.1
rho	–	0.472	0.472
Wald test – Chi-squared	4578	3444	3444
Log-likelihood	−7988	−18,629	−18,629
Pseudo R squared	0.261	–	–

Notes:
In column (1) the dependent variable is equal to one if a firm engaged in external R&D (R&D bought from other partners) and zero otherwise. In column (2) the dependent variable is equal to one if a firm engaged continuously in R&D. The dependent variable in column (3) is equal to the logarithm of the ratio of total R&D expenditures over sales. In columns (1) and (2) coefficients display marginal effects computed at the sample mean. All models include industry and country dummies.
Heteroskedasticity-robust standard errors in parentheses *** p < 0.01, ** p < 0.05, * p < 0.1.

results of a probit estimation in which the dependent variable is the dummy indicating whether or not a firm has invested in R&D performed outside the firm. The estimated coefficients of the probit estimation enable us to build a measure of the openness variable (O_i) through the predicted values of the dependent variable. We hence obtain a useful weight, bounded between zero and one, which we multiply by the amount of R&D in the sector of each firm in order to obtain a proxy of the horizontal flows of knowledge (*horizontal* O_i*).

In order to estimate equation (5.5) we also need to take into account the large number of firms that declare zero expenditure in internal R&D, which might lead to the usual problems of censored distribution. In columns (2) and (3) of Table 5.3 we implement a Tobit type II estimation procedure to identify the determinants of the intensity of R&D expenditures. This estimation allows us to obtain the predicted values of internal R&D intensity for all the firms in the sample as a proxy of the investments in internal knowledge (IK) in equation (5.1).

In Table 5.4 we estimate the main equation of interest – equation (5.5) – in which we test the effect of all types of flows of external knowledge on firms' innovative performance. In columns (1) and (2) the dependent variable is a dummy variable for the introduction of at least one product and/or process innovation.

The predicted R&D intensity is positive and significant, coherent with previous studies (Griffith et al., 2006a; Li-Ying et al., 2013; Voudouris et al., 2012). As for the vertical flows of knowledge, both downstream flows (*downstream*) and upstream flows (either transaction- or interaction-based) have a positive and significant effect on the probability of introducing an innovation, highlighting the important role of vertical ties. The variable that proxies the horizontal flows of external knowledge (*horizontal* O_i*) displays a positive and significant coefficient, suggesting that also the ability to benefit from competitors' stock of knowledge contributes positively to innovation performance. In column (2) we use the average level of intramural R&D performed in each sector, instead of the overall sum: the coefficient of horizontal flows increases, suggesting that average R&D is probably a more precise measure of the stock of horizontal external knowledge available to a firm.

In the remaining columns of Table 5.4 we distinguish between the introduction of product and process innovation. In line with our hypothesis *H1* we find that the horizontal flows of knowledge (*HORIZONTAL*) have a much larger positive effect on product innovation rather than on process innovation. When in columns (4) and (6) the horizontal flows of external knowledge are proxied by the average level of R&D in each sector (weighted by the degree of openness), rather than the simple sum, the

Table 5.4 Probit estimation of product and process innovation

Dependent variable	Product and/or process innovation		Product innovation		Process innovation	
	(1)	(2)	(3)	(4)	(5)	(6)
upstream-interaction	0.253***	0.255***	0.090***	0.092***	0.223***	**0.225***
	(0.016)	(0.016)	(0.013)	(0.013)	(0.012)	**(0.012)
upstream-transaction	0.087***	0.087***	0.042***	0.042***	0.054***	**0.054***
	(0.002)	(0.002)	(0.001)	(0.001)	(0.001)	**(0.001)
downstream	**0.305***	**0.306***	**0.279***	**0.280***	0.066***	0.068***
	(0.015)	(0.015)	**(0.013)	**(0.013)	(0.011)	(0.011)
horizontal*O_i (sum of sectoral R&D)	**0.037***		**0.039***			
	(0.003)		**(0.003)			
horizontal*O_i (average sectoral R&D)		0.048***		**0.052***	0.014***	0.014***
		(0.005)		**(0.004)	(0.002)	(0.002)
Predicted R&D intensity	0.497***	0.511***	0.349***	0.360***	0.173***	0.187***
	(0.019)	(0.019)	(0.014)	(0.014)	(0.010)	(0.010)
Turnover in 2004 (in logs)	0.136***	0.143***	0.090***	0.097***	0.049***	0.057***
	(0.006)	(0.006)	(0.005)	(0.005)	(0.004)	(0.004)
Observations	23,247	23,247	23,247	23,247	23,247	23,247
Log-likelihood	−8054	−8074	−9179	−9204	−9972	−9991
Pseudo R squared	0.498	0.497	0.396	0.394	0.299	0.297

Notes:
All models report marginal effects from a probit estimation. All models include country and sectoral dummies.
Heteroskedasticity-robust standard errors in parentheses *** p < 0.01, ** p < 0.05, * p < 0.1.

Table 5.5   Effects of different types of external knowledge on product and process innovation

Type of external knowledge	Product innovation	Process innovation
Horizontal flow	**Large**	Small
Downstream interaction	**Large**	Small
Upstream interaction	Small	**Large**
Upstream transaction	Small	**Large**

difference between product and process innovations becomes even more evident: in the product innovation equation the coefficient of horizontal flows of knowledge displays a higher coefficient than in the previous specification, while in the case of process innovation the effect of the horizontal flows of external knowledge becomes smaller.

The same differentiated effect is found for the external knowledge coming from the downstream vertical linkages (*downstream*), proxied by the importance of customers for the development of innovations, which tend to favour product innovation, rather than process innovation, in line with Hypothesis *H3*. Conversely, upstream vertical flows of knowledge have a large and positive impact on process innovation, coherently with Hypothesis *H2*. Both transaction-based upstream linkages (*upstream transaction*) proxied by the intensity of the expenditure in machinery acquisition, and interaction-based linkages (*upstream interaction*), proxied by the importance of suppliers, display a large and positive coefficient in the process innovation equation, while their impact is considerably smaller for product innovation (Table 5.5).

Heterogeneous effects
In Tables 5.6 and 5.7 we check if the results change for different subsamples of firms: we run separate regressions for each size category of firm, distinguishing between firms with up to 50 employees, firms with 50 to 250 employees and large companies with more than 250 employees. Table 5.6 shows that the results are extremely robust across the three samples. Transaction-based and interaction-based upstream vertical flows of knowledge are always more important for the introduction of process innovations, while downstream vertical interactions with customers always show a larger effect for product innovation among small, medium and large firms. The only difference with respect to the baseline specification consists in the coefficient of horizontal flows for large firms, which becomes almost the same both for product and process innovations, suggesting that large firms benefit less from competitors' knowledge for the introduction of product innovations.

Table 5.6 *Probit estimation of product and process innovation: different firm size*

Variables	(1)	(2)	(3)	(4)	(5)	(6)
	Small firms		Medium-sized firms		Large firms	
	Product inno.	Process inno.	Product inno.	Process inno.	Product inno.	Process inno.
upstream-interaction	0.076***	**0.170***	0.087***	**0.228***	0.058**	**0.260***
	(0.016)	**(0.017)**	(0.023)	**(0.021)**	(0.024)	**(0.025)**
upstream-transaction	0.033***	**0.045***	0.037***	**0.059***	0.037***	**0.047***
	(0.001)	**(0.001)**	(0.002)	**(0.002)**	(0.003)	**(0.004)**
downstream	**0.253***	0.048***	**0.261***	0.064***	**0.172***	0.054**
	(0.020)	(0.014)	**(0.020)**	(0.018)	**(0.020)**	(0.024)
horizontalO_i *(average sectoral R&D)*	**0.055***	0.019***	**0.060***	0.032***	**0.049***	**0.043***
	(0.005)	(0.004)	**(0.006)**	(0.005)	**(0.006)**	**(0.006)**
Predicted R&D intensity	0.280***	0.175***	0.408***	0.201***	0.281***	0.144***
	(0.017)	(0.013)	(0.025)	(0.020)	(0.027)	(0.026)
Turnover in 2004 (in logs)	0.079***	0.061***	0.107***	0.049***	0.067***	−0.012
	(0.006)	(0.005)	(0.011)	(0.009)	(0.013)	(0.013)
Observations	12,841	12,841	7133	7133	3273	3273
Log-likelihood	−4492	−4515	−3218	−3415	−1315	−1819
Pseudo R squared	0.379	0.341	0.338	0.264	0.381	0.197

Notes:
All models report marginal effects from a probit estimation. All models include country and sectoral dummies.
Heteroskedasticity-robust standard errors in parentheses *** $p < 0.01$, ** $p < 0.05$, * $p < 0.1$.

Table 5.7   Probit estimation of product and process innovation: high-tech and low-tech sectors

Dependent variable	(1)	(2)	(3)	(4)
	High-tech		Low-tech	
	Product inno.	Process inno.	Product inno.	Process inno.
upstream-interaction	0.079***	**0.215***	0.074***	**0.197***
	(0.023)	**(0.020)**	(0.014)	**(0.015)**
upstream-transaction	0.046***	**0.048***	0.034***	**0.056***
	(0.002)	**(0.002)**	(0.001)	**(0.001)**
downstream	**0.284***	0.038**	**0.241***	0.092***
	(0.017)	(0.015)	**(0.017)**	(0.015)
horizontalO_i (average sectoral R&D)*	**0.053***	0.026***	**0.061***	0.026***
	(0.005)	(0.004)	**(0.005)**	(0.004)
Predicted R&D intensity	0.351***	0.119***	0.295***	0.232***
	(0.023)	(0.016)	(0.015)	(0.014)
Turnover in 2004 (in logs)	0.079***	0.035***	0.076***	0.060***
	(0.009)	(0.006)	(0.005)	(0.005)
Observations	7740	7740	15,507	15,507
Log-likelihood	−3307	−3866	−5804	−5920
Pseudo R squared	0.383	0.220	0.368	0.358

Notes:
All models report marginal effects from a probit estimation. All models include country and sectoral dummies. High-tech sectors include manufacture of: chemicals and chemical products (ISIC 24); machinery and equipment (ISIC 29); office machinery and computers (ISIC 30); electrical machinery and apparatus n.e.c. (ISIC 31); medical, precision and optical instruments (ISIC 33); and motor vehicles and other transport equipment (ISIC 34 and 35). Heteroskedasticity-robust standard errors in parentheses *** $p < 0.01$, ** $p < 0.05$, * $p < 0.1$.

Finally, we control whether our results change if we consider firms active in sectors with different levels of technological intensity: specifically, we want to check whether firms in low- or high-tech sectors benefit differently from the heterogeneous sources of external knowledge that we have identified. In Table 5.7 we split our sample between high-tech sectors and low-tech sectors. Again our results are very robust: in both subgroups product innovations benefit mainly from horizontal flows of knowledge and downstream vertical linkages, while upstream vertical knowledge flows are more relevant for the introduction of process innovations.

4. IMPLICATIONS AND CONCLUSIONS

The successful introduction of technological innovations is the result of the matching of characteristics of the system in which firms are embedded and their strategies. Technological change is the result of the interdependence between the action of individual agents and the structural characteristics of the system. It occurs when and if the structure of the system in which firms are localized provides access to external knowledge enabling the creative reaction of firms that leads to the introduction of innovations.

Knowledge, in general, and external knowledge specifically, is a bundle of different types of knowledge that it seems necessary to disentangle. This requires the twin effort to specify: i) the types of external knowledge according to source, whether horizontal (that is, within the same industry) or vertical (across the chains of user–producer interactions-cum-transactions), distinguishing between downstream (with customers) and upstream (with suppliers) interactions; and ii) their effects on the types of innovations introduced, whether product or process.

The inbound search for the proper match between types of external knowledge and their relative costs and innovation strategies has important implications for the strategies of firms. Firms will take into account the important differences in the actual access conditions and absorption costs of the different types of external knowledge in the selection of their innovation strategies.

We put forward the hypotheses that horizontal flows of knowledge should induce the introduction of product innovations because of the borrowing and imitation of knowledge from competitors engaged in the same product-innovation races. Also, downstream flows of knowledge from customers should favour product innovations, thanks to their competent knowledge of the products sold by the innovating firm. In contrast, upstream sources of knowledge from suppliers, either based on the purchasing of advanced machineries or in true interactions and collaborations, should lead to process innovations.

In our empirical analysis we explore the direct effects of the access conditions to the different types of external knowledge on the actual amount of product and process innovations introduced by European firms in 2004, using CIS data. The dependent variable of our analysis is the introduction of process and product innovations. In order to assess their role, the horizontal flows of external knowledge, as proxied by the flows of R&D activities performed by competitors (firms active in the same industry), have been filtered by the degree of openness to external knowledge as measured by the firm-specific propensity to invest in extramural R&D. Vertical flows of external knowledge have been distinguished between upstream and downstream user–producer interactions. In order to grasp the role of knowledge interactions we utilized the information available in the CIS data based on the perceived relevance of external vertical sources of knowledge, whether stemming from interactions with customers or suppliers. Finally, in order to appreciate the amount of external knowledge flowing via user–producer transactions we have taken into account the intensity of investment as a proxy of machinery purchased by downstream users.

The results confirm our hypotheses: the upstream vertical flows of external knowledge channelled both by transactions and interactions have positive effects mainly on the introduction of process innovations. The downstream vertical flows of external knowledge, as well as the flows of external knowledge spilling horizontally from competitors, have positive effects mainly on the introduction of product innovations. Moreover, the findings are surprisingly similar for different types of firm: even when we consider separately small, medium and large firms, as well as high-tech and low-tech ones, our results remain largely unchanged.

Our analysis provides an important contribution to the related literature on inbound open innovation practices (Chesbrough and Crowther, 2006). This literature has found relevant limits and drawbacks of the quantitative effect of external knowledge on innovation outcomes (Howells et al., 2008; Grimpe and Kaiser, 2010). However our results show that an appropriate analysis of the effect of external knowledge on innovation must account for the degree of heterogeneity of external knowledge, as well as its differentiated impact on different innovative strategies. Indeed, the results stress the diversity of the effects of knowledge externalities, and confirm the interest to appreciate the variety of types of external knowledge and the need to explore in detail their different sources and differentiated effects.

The distinction of external knowledge according to its origin and the identification of its differentiated effects favouring different innovation strategies also represent a step forward in the identification of the effects of the system upon the overall innovation process. The necessary corollary of the analysis is in fact that not only the amount of innovations being introduced but also their typology is influenced by the characteristics of the system in which firms are embedded.

These results seem quite important for R&D management for their strategic implications for inbound exploratory practices. Firms that are willing to rely on the introduction of process innovations in product markets characterized by cost and price competition should pay attention to the opportunities provided by vertical transactions and interactions with their suppliers. Firms that are more engaged in product innovation races within oligopolistic markets should pay attention to the opportunities provided by both customers and competitors as sources of dedicated knowledge externalities.

The careful identification of the role of innovation in the strategy of the firm, the selection of the types of innovation (whether product or process) and the screening of the types of technological spillover available in an economic system should be part of a single and integrated decision-making process. The availability of major vertical flows of external knowledge provided by upstream providers of capital goods and intermediary inputs might favour the choice of introducing process rather than product innovations. The search for the accurate matching between types of innovation and types of external knowledge should become an important procedural device for decision-making.

Our empirical analysis has some limitations that should be highlighted. Our measures of the importance of the different sources of knowledge are simple dummy variables that might represent a quite poor proxy of the level of importance of each flow of knowledge. Moreover, the debate on technology-push and demand-pull has shown that among the push-factors universities and PROs are crucial sources of science-based external knowledge (Di Stefano et al., 2012), but in our analysis we did not include them among the possible sources of external knowledge. Further research could hence improve on such shortcomings of our study.

Our analysis on the heterogeneity of external knowledge calls for further research on the relationships between sources of knowledge and innovative strategies. A first possible extension is to better distinguish between inbound open innovation practices based on transactions, as opposed to those that are based on interactions, whether vertical upstream or downstream. Moreover, in this study we only analysed the inbound side of open-innovation practices: however to every inbound flow in a system should correspond an outbound flow (Chesbrough and Crowther, 2006; Knudsen and Mortensen, 2011). Therefore the necessary corollary of our analysis is to check also the effect of outbound strategies on the direction of firms' innovations.

NOTES

1. The authors acknowledge the provision of the European Commission, Eurostat, Fourth Community Innovation Survey, microdata. Eurostat has no responsibility for the opinions expressed in this chapter, which remain solely of the authors. Eurostat releases data in micro-aggregated form for confidentiality reasons.
2. We use the same procedure adopted by Hall and Mairesse (1995) and remove any observations for which value added in 2002 or 2004 is zero. We also eliminate any observations for which the growth rate of value added is less than minus 90 per cent or greater than 300 per cent. Finally, we drop firms for which the ratio between total R&D expenditures and value added is higher than 80 per cent.

6. The cost of knowledge

Cristiano Antonelli and Alessandra Colombelli*

1. INTRODUCTION

The study of the cost of knowledge seems an important area of investigation that has received, so far, quite surprisingly, very little attention. After the introduction of the knowledge generation function it is now necessary to introduce and analyse the knowledge cost function.

Identification of the knowledge generation function has been a major advance in the economics of knowledge (Crépon et al., 1998; Weitzman, 1998). Finally, technological knowledge can be analysed as the output of a dedicated economic activity. Working along these lines, increasing evidence shows that the unit costs of knowledge differ widely across firms. Some firms are able to generate new technological knowledge with low levels of current expenditures in R&D. Others experience very high levels of current expenditures. As a matter of fact the costs of knowledge differ and their variance becomes a fascinating area of research. The new appreciation of the role of knowledge indivisibility in the generation of new knowledge enables us to better grasp the specific effects of knowledge externalities and knowledge cumulability on the costs of knowledge (Antonelli and Colombelli, 2015a, b).

The rest of the chapter is structured as follows. Section 2 recalls the recent advances of the new economics of knowledge and applies them to grasping the determinants of the heterogeneity of firms in terms of unit costs of their knowledge. Section 3 provides an empirical investigation of the econometric estimate of a knowledge cost function based on a panel of companies listed on the financial markets in the UK, Germany, France, Italy and the Netherlands for the period 1995–2006 for which information about patents has been gathered. The conclusions summarize the results and discuss the implications of the analysis.

2. KNOWLEDGE AS INPUT AND OUTPUT

After a long period of time during which the early economics of knowledge has investigated in depth the determinants and effects of the characteristics of knowledge as a good – with special attention to its limited appropriability, non-rivalry in use and non-tradeability – the new economics of knowledge pays much attention to the characteristics of the knowledge generation process. In this context, it has grasped the implications of another bundle of characteristics of knowledge as an economic good that has received less attention: knowledge indivisibility, in terms of both knowledge cumulability and knowledge complementarity. The twin character of knowledge – at the same time an input and an output – and its limited exhaustibility enable us to grasp a key aspect of the knowledge generation process. Its generation consists in the recombination of knowledge items that enter the process as inputs (Weitzman, 1996, 1998).

Because of knowledge complementarity and knowledge cumulability, together with current R&D activities, both the external knowledge generated by third parties but not fully appropriated and the internal stocks of knowledge generated by each firm in the past are now recognized as relevant inputs in the generation of knowledge as an output. The knowledge generated as the output of a dedicated activity is itself a necessary condition and hence an input for both the introduction of an innovation and the generation of further knowledge (David, 1993). This has led to analysis of the generation of technological knowledge as a specific economic activity (Crépon et al., 1998; Lööf and Heshmati, 2002; Nesta and Saviotti, 2005).

A second important step in this enquiry can be done with analysis of the knowledge cost function. This approach enables us to identify the determinants of the great variance in the costs of knowledge. Specifically, study of the knowledge cost function helps with grasping to what extent the cost of knowledge is affected by the availability of the full range of inputs and their costs (Antonelli and David, 2016).

As soon as it becomes evident that R&D activities are not the single input in the knowledge generation process (Gunday et al., 2011), the stocks and composition of existing knowledge both internal and external to each firm, as indispensable and strictly complementary inputs, acquire a new relevance (Antonelli and Colombelli, 2015a, b). Knowledge inputs such as the amount of external knowledge that can be accessed by firms to generate new knowledge are distributed unevenly across space. Major institutional and structural characteristics affect the actual amount of external knowledge that each firm can access and use as an input. The costs of these inputs differ in turn because of the variance in the access conditions to the external knowledge available (Cohen and Levinthal, 1989, 1990) and because of the different characteristics of the local pools of external knowledge (Saviotti, 2007; Quatraro, 2010, 2012). By the same token, firms differ widely with respect to the size and characteristics of the stocks of internal knowledge that can be used to generate new knowledge (Jones, 1995). Knowledge inputs and outputs vary across firms also because firms differ in their specific competence in managing the knowledge generation process (Nelson, 1982).

The inclusion in the knowledge cost function of these variables stems from the identification of the recombinatorial character of the knowledge generation process, and enables us to appreciate the role of knowledge indivisibility as articulated in knowledge cumulability and knowledge complementarity in its generation (Weitzman, 1996, 1998). Let us consider these aspects in turn.

Knowledge cumulability – and its limited exhaustibility – implies that the stock of existing knowledge can be used again and again and plays a central role as input in the generation of new knowledge. The stock of knowledge qualifies and identifies the knowledge base of each firm. The inclusion of this variable enables us to grasp the path-dependent character of knowledge generation. The generation of new technological knowledge at each point in time, by each agent, in fact is strongly influenced by the accumulation of knowledge in the past. The current levels of R&D expenditures of each agent do play a role, but only in a context that is shaped by the past of each firm (Antonelli, 2011; Belenzon, 2012).

Appreciation of knowledge complementarity enables us to put in context the role of knowledge externalities. A large body of literature has explored the role of technological spillovers as a major input in the generation of new technological knowledge (Colombelli et al., 2013b). In this approach external knowledge plays an important and yet supplementary role in the generation of new technological knowledge (Griliches, 1979, 1990, 1992). Moreover, its recipients are mainly viewed as the passive beneficiaries of knowledge leaking from other firms (Feldman, 1999). A large body of empirical evidence has subsequently confirmed that external knowledge is an essential input in the generation of new knowledge (Adams, 1990; Marrocu et al., 2012; Smit et al., 2015).

The composition of the knowledge pools to which co-localized firms have access also plays an important role in assessing the levels of absorption activities (Camagni and Capello, 2013; Grillitsch et al., 2013). Technological knowledge cannot be regarded as a homogeneous pile, but rather as a composite bundle of highly differentiated and idiosyncratic elements that are qualified by specific relations of interdependence and interoperability. This approach enables us to identify the extent to which the generation of new technological knowledge in a field depends upon the contributions of knowledge inputs stemming from other fields: a new knowledge item exhibits high levels of compositeness when it relies upon a large number of other knowledge fields (Antonelli, 2011). The quality of the local pools of knowledge, in other words, matters as well as its sheer size. The larger the coherence of the local knowledge base and the shorter the distance between different types of knowledge, the higher the probability that they can be combined (Saviotti, 2004, 2007; Krafft et al., 2009; Quatraro, 2010).

The interplay between the stock and composition of internal knowledge, which also increases a firm's absorptive capacity, and the actual levels of knowledge externalities helps to increase the amount of knowledge that each firm can generate with a given amount of R&D activities and competence acquired by means of internal learning processes.

Analysis of a knowledge cost function that takes into account the role of internal stocks of knowledge and of local pools of external knowledge enables us to reconsider, but from quite a different perspective, two standard assumptions of the economics of innovations – that is, the well-known Schumpeterian and Marshallian hypotheses. Let us consider them in turn.

2.1 The Schumpeterian Hypothesis

Joel Mokyr (1990b: 267) has recently expertly summarized the Schumpeterian hypothesis as follows: 'large firms with considerable market power, rather than perfectly competitive firms are the "most powerful engine of technological progress"'. Schumpeter, in *Capitalism, Socialism and Democracy*, actually went so far as to claim that perfect competition is not only impossible but also 'inferior' (Schumpeter, 1942: 106). The Schumpeterian hypothesis has fed a long-lasting theoretical debate, and the significant empirical evidence provided controversial proof of the actual advantages of large firms with respect to smaller ones in the rates of generation of technological knowledge and the eventual introduction of innovations. The results of empirical studies in different sectors, historic periods, countries and regions have not provided conclusive evidence (Link, 1980; Link and Siegel, 2007). The recent advances in the economics of knowledge enable us to focus the Schumpeterian hypotheses on knowledge generation and on the long-lasting effects of the limited divisibility and exhaustibility of knowledge. The Schumpeterian hypothesis, in other words, would apply only to the size of the stock of knowledge and not to the sheer size of firms in terms of employment.

Following this approach, we put forward the specific hypothesis that the size of firms exerts negative – cost-reducing – effects when it is measured in terms of internal knowledge stock rather than in terms of sheer size. For a given size in terms of employment, firms with a larger stock of internal knowledge have lower unit knowledge costs than firms with smaller internal stocks. The advantage of incumbents, in other words, stems specifically from the effects of knowledge cumulability and non-exhaustibility, and is specific to the size of the stock of knowledge.

2.2   The Marshallian Hypothesis

According to the standard application in the economics of knowledge of the Marshallian hypothesis, firms located in large industrial districts with a strong knowledge base have better chances to access knowledge spillovers and feed their own knowledge generation process. In large districts with a rich knowledge base firms have better access to external knowledge and can substitute it with expensive R&D activities. Knowledge externalities are pecuniary rather than pure: relevant search and absorption costs are necessary in order to use knowledge spilling from third parties as an input in the generation of new knowledge. For this reason, not only the size of the knowledge base but also its nature is of some significance. Indeed, the generation of technological knowledge stems from a variety of competences: knowledge is not a homogeneous good, and therefore its intrinsic heterogeneous nature cannot be neglected.

We thus argue that not only the amount of knowledge available at the local level but also the characteristics of that knowledge have an impact on the costs of knowledge. More precisely, the larger is the size and coherence of the local knowledge pools and its complementarity with the internal knowledge base, the lower the search costs (Antonelli, 2008). On the other hand, the higher the variety and the dissimilarity in the combination of technologies in the firm's region, the higher the costs associated with the firm's knowledge output. When the local knowledge that firms may access is distributed across a wide range of technology domains and features technologies which are far away from one another in the technological space, the absorption costs increase.

Following these arguments along the lines of the Marshallian tradition of analysis, we put forward two hypotheses: i) we expect a negative correlation between the size of the regional stock of knowledge and the firm cost of knowledge; ii) knowledge externalities are all the more effective the larger are the levels of coherence and the lower the levels of variety and dissimilarity of the local knowledge base. We thus expect that the unit knowledge costs decrease with regional levels of knowledge coherence and increase with regional levels of variety and dissimilarity.

The following knowledge cost function provides the general framework of our approach:

$$CK_{it} = (R\&D_{it} KNOWLEDGEBASE_{it} EXTERNAL\ KNOWLEDGE_{it}) (6.1)$$

Equation (6.1) provides a suitable specification of the knowledge cost function that accommodates, along with the role of R&D expenditures, appreciation of the knowledge base of each firm in terms of stocks of knowledge in the generation of new knowledge, and identification of the key role of knowledge external to each firm but available in regional proximity. Specifically, we expect that unit knowledge costs are lower the larger the stock of internal knowledge and the larger the pool of external knowledge that firms can access, as well as its consistency with the stock of internal knowledge.

3. EMPIRICAL EVIDENCE

3.1 Dataset

As in Chapter 4, our data source is the IPER database, which collects information on 3382 active companies listed in the UK, Germany, France, Italy and the Netherlands.[1] These countries were selected not only for their economic size and importance but also as they represent the main European financial markets. The variety in terms of size, sectors, regions and countries of this set of companies seems to provide a reliable representation of the European business sector. The IPER database has been built by matching information from multiple sources of data. Our main source of market and accounting data is Thomson Reuters Datastream, which delivers worldwide economic and financial time series data. To obtain additional relevant variables, we again include information from Bureau Van Dijk's Amadeus. In order to match information from these two databases, we used International Securities Identification Numbers (ISIN).

We also use data from the OECD REGPAT database on patent applicants and inventors as well as on technological classes cited in patents granted by the European Patent Office (EPO) and the World Intellectual Property Organization (WIPO), under the Patent Cooperation Treaty (PCT), from 1978 to 2006. The use of patents as the single indicator of the knowledge output is, indeed, a limit of the analysis. A large body of literature has identified the limits of patents as the exclusive source of information on the actual amount of knowledge generated: not all firms patent their 'inventions'; small firms rely less than large ones on patents to increase the appropriability of their inventions; patents are used more to secure property rights of inventions that apply to product rather than process inventions; firms in fashion industries rarely patent their distinctive knowledge. Awareness of these limits has not prevented the use

Table 6.1   Sample distribution in macro-sectors

Macro-sector	%	Cum.
High-technology manufactures – HT	31.6	31.6
Medium/high-technology manufactures – MHT	45.4	77.0
Medium/low-technology manufactures – MLT	4.5	81.5
Low-technology manufactures – LT	8.9	90.4
Knowledge intensive sectors – KIS	9.4	99.8
Less knowledge-intensive sectors – LKIS	0.2	100.0
Total	100.0	

of patents in the extensive empirical literature that relies on the legacy of Zvi Griliches (1984, 1990).

In order to match the firm-level data with data on patents, we draw on the work of Thoma et al. (2010), which develops a method for harmonization and combination of large-scale patent and trademark datasets with other sources of data, through standardization of applicant and inventor names. The new evidence on the actual meaning of patent citations, often included by patent officers to better specify the borders of the domain of the intellectual property right, rather than its quality, suggests using the raw evidence of the number of patents with no attempt to try to elaborate misleading quality indicators (Van Zeebroeck, 2011; Van Zeebroeck and van Pottelsberghe, 2011).

Finally, we pooled the dataset by adding industry-level information from the OECD STAN databases as it is based on ISIC revision 3 sectoral classifications and Datastream uses the four-digit level ICB industry classification (see Antonelli and Colombelli, 2015a, b).

Our final dataset includes active companies listed on the main European financial markets that submitted at least one patent application to the EPO in the period analysed. Table 6.1 reports the sample distribution by macro-sector. High- and medium/high-technology firms account for around 31.6 per cent and 45.4 per cent of observations, respectively. Medium/low and low-tech firms account for 4.5 per cent and 8.9 per cent respectively, while knowledge-intensive firms represent 9.4 per cent of observations.

3.2   The Econometric Analysis: Methodology and Variables

The econometric analysis is organized on a baseline equation and a number of complementary specifications that explore in detail the different facets of the basic hypotheses. Our baseline estimating equation is:

$$PCost_{it} = \beta_1 + \beta_2 R\&D_{it-1} + \beta_3 PStock_{it-1} + \beta_4 AGE_{it-1} + \beta_5 Reg PStock_{it-1} +$$
$$\beta_6 RegTV_{it-1} + \beta_7 RegCD_{it-1} + \beta_8 RegCOH_{it-1} + \sum \rho_i + \sum \psi_t + \varepsilon_{it} \quad (6.2)$$

In equation (6.2) all explanatory variables are lagged one year so as to mitigate endogeneity problems. Given the panel nature of our dataset and to control for unobserved firm-specific characteristics, equation (6.2) has been estimated using a fixed effects estimator. The Hausman test confirms that fixed effects perform better than the random effects estimator.

In equation (6.2) the dependent variable for the firm i at time t is the cost of knowledge output measured by the logarithm of the ratio between the firm's current R&D expenditures and the number of patents delivered. This measure is a good proxy of the actual cost of producing new technological knowledge. Yet, it is worth noting that the cost of external interactions based on knowledge is not directly accounted for. The unit cost of knowledge is explained by two sets of independent variables: i) the knowledge base of each firm as defined by the internal expenses in R&D, the size of the internal knowledge stock and the age of the firm; and ii) the size the local pools of external knowledge and their composition in terms of variety, complementarity and similarity (dissimilarity). As to the latter set of variables, variety aims to capture the technological differentiation within the knowledge base of a region; coherence measures the extent to which the pieces of knowledge that firms combine to generate new technological knowledge are complementary; and, finally, similarity (dissimilarity) measures the extent to which the pieces of knowledge used by firms are close (distant) to one another in the technology space (Krafft et al., 2014).

More precisely, on the right-hand side, the first set of variables considers *R&D*, that is, the current research efforts and activities funded by each firm at time $t-1$, measured as the ratio of R&D expenditures (*R&Dexp*) to total assets (in logarithms). In order to appreciate the effects of the stocks of internal knowledge of firms, the model includes the variable *PStock* measured as the ratio between the number of patents held by each firm (*CumPatStock*) and total assets (in logarithms). *CumPatStock* is computed by applying the permanent inventory method (PIM) to patent applications. We calculate it as the cumulated stock of patent applications using a rate of obsolescence of 15 per cent per annum:[2]

$$CumPatStock_{it-1} = \dot{h}_{it-1} + (1 - \sigma)\, CumPatStock_{it-2} \quad (6.3)$$

where $\dot{h}_{it-1}$ is the flow of patent applications (in logarithms) and σ is the rate of obsolescence. Alongside the stock of patents we include the age of firms. This variable aims at grasping the effects of the accumulation of competence by means of learning processes. *Age* is measured by the years since foundation and is expressed in logarithms; it aims at grasping the effects of the accumulation of tacit knowledge and competence, based upon learning processes that affect the generation of patentable knowledge.

To articulate the different facets of the knowledge that is external to each firm and made accessible by proximity with firms co-localized, in the second basket of variables we include first a variable aimed at grasping the effects of the size of the knowledge pools in which firms are embedded: *RegPStock*, that is the log of patents stock in the same region (NUTS2) of firm i at time $t - 1$. The method used for computing this variable is the same used for *PStock*, that is, (PIM).

Next, we include other variables proxying for variety, complementarity and similarity. These indicators rest on the recombinant knowledge approach. In order to provide an operational translation of such concepts one needs to identify both a proxy for the bits of knowledge and a proxy for the elements that make up their structure. We consider patents as a proxy for knowledge, and then look at technological classes to which patents are assigned as the constituting elements of its structure, that is, the nodes of the network representation of recombinant knowledge. Each technological class j is linked to another class m when the same patent is assigned to both of them. The higher the number of patents jointly assigned to classes j and m, the stronger is this link. Since technological classes attributed to patents are reported in the patent document, we will refer to the link between j and m as the co-occurrence of both of them within the same patent document. On this basis we calculated the following three key characteristics of firms' knowledge bases (all expressed in logarithms):

- Knowledge variety (KV) measures the degree of technological diversification of the knowledge base. It is based on the informational entropy index. We thus include in equation (6.2) *RegTV* as a measure of the regional total variety, *RegRTV* and *RegUTV*, measuring the related and unrelated variety respectively (see Antonelli and Colombelli, 2015a, b for methodological details). Unrelated variety measures the technological diversification of the knowledge base which is likely to be affected by a radically new type of knowledge, while related variety measures the technological diversification of the knowledge base which is likely to be affected by incremental recombination of existing types of knowledge.
- Knowledge coherence (COH) measures the degree of complementarity among technologies. It is measured by means of the *RegCOH* index (see Antonelli and Colombelli, 2015a, b).

- Cognitive distance (CD) expresses the dissimilarities among different types of knowledge and is measured using the *RegCD* variable (see Antonelli and Colombelli, 2015a, b).

The inclusion of these variables marks an important step forward in the operational translation of knowledge creation processes. In particular, they allow for a better appreciation of the collective dimension of knowledge dynamics. Knowledge is indeed viewed as the outcome of a combinatorial activity in which intentional and unintentional exchange among innovating agents provides access to external knowledge inputs (Fleming, 2001; Fleming and Sorenson, 2001).

The recombinant knowledge approach provides indeed a framework to represent the internal structure of regional knowledge bases as well as to enquire into the effects of their evolution (Antonelli et al., 2010). If knowledge stems from the combination of different technologies, knowledge structure can be represented as a web of connected elements. The nodes of this network stand for the elements of the knowledge space that may be combined with one another, while the links represent their actual combinations. The frequency with which two technologies are combined provides useful information on the basis of which one can characterize the internal structure of the knowledge base according to the average degree of complementarity and proximity of the technologies which knowledge bases are made of, as well as to the variety of the observed pairs of technologies.

The dynamics of technological knowledge can therefore be understood as the patterns of change in its own internal structure, that is, in the patterns of recombination across the elements in the knowledge space. This allows for qualifying both the cumulative character of knowledge creation and the key role played by the properties describing knowledge structure (Saviotti, 2004, 2007; Quatraro, 2010; Colombelli et al., 2013b). Finally, we include time dummies in order to control for time effects.

In order to check further the robustness of our empirical analysis with respect to the role of external knowledge, we also estimated an extended model including the patenting activities of firms outside a firm's region (*WRegPStock*). Here *WRegPStock* aims at capturing the role of the sources of external knowledge that are far away from firm i. The variable *WRegPStock* has been computed as the log of patents stock (PIM) in the NUTS2 regions of the EU-24 member states, weighted using a row-normalized inverse distance matrix so as to appreciate the contribution of knowledge produced in regions close to firm i's region at time $t - 1$. Moreover, as a further robustness check, we also estimated additional models including firm *Size* among the covariates. The inclusion of a variable that accounts for the sheer size (that is, the logarithm of employees

Table 6.2   Variables' measurement methods

Variable	Measurement method
PCOST	Log (R&D/N Patents) for firm i at time t
R&D	Log (R&D/Total assets) for firm i at time $t - 1$
R&Dexp	Log (R&D) for firm i at time $t - 1$
PStock	Log of (Patents stock (PIM)/Total assets) for firm i at time $t - 1$
CumPatStock	Log of (Patents stock (PIM)) for firm i at time $t - 1$
Age	Log of years since foundation for firm i at time $t - 1$
Size	Log of employees number for firm i at time $t - 1$
RegPStock	Log of patents stock (PIM) in the same region (NUTS2) of firm i at time $t - 1$
WRegPStock	Log of patents stock (PIM) belonging to EU-24 member states other than that of firm i at time $t - 1$, weighted using a row-normalized inverse distance matrix
RegTV	Log of total variety in the region (NUTS2) of firm i at time $t - 1$
RegRTV	Log of related variety in the region (NUTS2) of firm i at time $t - 1$
RegUTV	Log of unrelated variety in the region (NUTS2) of firm i at time $t - 1$
RegCD	Log of cognitive distance in the region (NUTS2) of firm i at time $t - 1$
RegCOH	Log of knowledge coherence in the region (NUTS2) of firm i at time $t - 1$

number for firm i at time $t - 1$) enables us to appreciate the estimated parameters as the direct effect of the variables proxying for the internal knowledge base, after taking into account the effects of the size of the firm.

For each variable the measurement method is defined in Table 6.2, while descriptive statistics are reported in Table 6.3. The correlation matrix for the extended model can be found in Table 6.4. As reported in the tables, correlations among some independent variables are relatively high. In particular, *RegPStock* is highly correlated with the three knowledge variety measures. Not surprisingly, the different measures of knowledge variety are highly correlated with each other. To further detect multicollinearity among covariates, we also checked the variance-inflation factor (VIF) for each covariate. If *RegPStock* is regressed on all the other covariates, including each of the three knowledge variety measures in different regressions, the VIF assumes values in the range 1.35–1.38, much less than the accepted cut-off value of 10 (Neter et al., 1990). Finally, when *RegPStock* is regressed on all the other covariates, including both *RegRTV* and *RegUTV*, the VIF value equals 1.43. Therefore, in our empirical analysis

Table 6.3   Descriptive statistics

Variable	Obs.	Mean	Std Dev.	Min.	Max.
PCOST	870	9.330	1.725	2.996	15.547
R&D	870	−3.340	1.097	−7.777	0.420
R&Dexp	870	10.888	2.146	2.996	15.824
PStock	870	−11.247	1.914	−18.732	−6.587
CumPatStock	870	2.981	1.756	−0.650	7.519
Age	870	3.425	1.185	0	5.541
Size	854	8.920	2.296	1.386	13.090
RegPStock	870	8.845	1.347	4.853	10.892
WRegPStock	870	7.627	0.268	6.988	8.268
RegTV	870	2.182	0.131	1.653	2.397
RegRTV	870	1.882	0.155	1.232	2.129
RegUTV	870	0.822	0.110	0.269	0.991
RegCD	870	−0.264	0.020	−0.368	−0.223
RegCOH	870	1.782	0.546	0.660	3.846

we ran different regression models. First, the three specifications of knowledge variety are included in different regression models. Subsequently, we include the two components of knowledge variety (*RegRTV* and *RegUTV*) in the same model. Moreover, we ran different regression models excluding knowledge stock from the vector of covariates. Finally, a relatively high correlation is also observed between internal R&D and the stock of patents. However, if *R&D* and *PStock* are regressed on all the other covariates the VIF assumes values in the range 2.61–2.84, much less than the cut-off value of 10. Therefore, we also ran different regression models excluding *R&D* from the vector of covariates as a robustness check.

3.3   Results

The results of the fixed effects regression estimations for equation (6.2) are reported in Table 6.5. The Hausman test, comparing the results obtained with the fixed effects model with those obtained from the random effects regression model, indicates that the fixed effects model is a better fit for our regressions. In order to cope with multicollinearity among the knowledge-related variables, column 1 shows the results for the baseline equation that only includes variables measuring the internal activities performed by each firm in terms of R&D expenditure, patent stock and age. Columns 2 to 5 include also the variables proxying for the size and composition of the external pool of knowledge. More precisely, the results of the model including the *RegTV* variable are presented in column 2.

Table 6.4 Correlation matrix

	PCOST	R&D	PStock	Age	RegPStock	WRegPStock	RegTV	RegRTV	RegUTV	RegCD	RegCOH
PCOST	1.000										
R&D	0.041	1.000									
PStock	−0.628	0.493	1.000								
Age	0.126	−0.207	−0.188	1.000							
RegPStock	0.224	−0.063	−0.162	0.034	1.000						
WRegPStock	0.025	0.033	0.215	−0.095	−0.045	1.000					
RegTV	0.149	0.019	−0.107	−0.020	0.821	0.025	1.000				
RegRTV	0.149	−0.002	−0.119	−0.015	0.814	0.056	0.982	1.000			
RegUTV	0.097	0.098	−0.014	−0.033	0.513	−0.116	0.660	0.509	1.000		
RegCD	0.175	−0.159	−0.064	0.010	−0.039	0.461	−0.153	−0.118	−0.224	1.000	
RegCOH	0.084	0.052	−0.023	−0.029	0.288	−0.267	−0.125	−0.146	0.019	−0.120	1.000

Table 6.5   Results: baseline model

Fixed effects	(1)	(2)	(3)	(4)	(5)
Variables	*PCost*	*PCost*	*PCost*	*PCost*	*PCost*
R&D	0.211***	0.197**	0.208***	0.196**	0.190**
	(0.0763)	(0.0768)	(0.0766)	(0.0769)	(0.0770)
PStock	−0.489***	−0.463***	−0.463***	−0.471***	−0.468***
	(0.0605)	(0.0613)	(0.0615)	(0.0614)	(0.0614)
Age	0.0145	0.0971	0.0787	0.0985	0.110
	(0.160)	(0.163)	(0.163)	(0.163)	(0.163)
RegPStock		−0.631*	−0.557	−0.331	−0.496
		(0.374)	(0.379)	(0.356)	(0.379)
RegTV		2.512**			
		(1.152)			
RegRTV			1.306		1.151
			(0.920)		(0.921)
RegUTV				1.837**	1.745**
				(0.866)	(0.869)
RegCD		8.910**	8.625**	9.248**	9.236**
		(4.156)	(4.162)	(4.166)	(4.164)
RegCOH		0.275	0.203	0.0814	0.216
		(0.238)	(0.240)	(0.214)	(0.240)
Constant	6.951***	8.953**	11.45***	10.40***	9.602**
	(0.881)	(3.987)	(3.722)	(3.773)	(3.826)
Observations	870	870	870	870	870
R-squared	0.386	0.395	0.393	0.395	0.396
Number of id	171	171	171	171	171

Notes:
Standard errors in parentheses.
*** $p < 0.01$, ** $p < 0.05$, * $p < 0.1$.

Columns 3 and 4 show the results for the *RegRTV* and *RegUTV* variables, respectively, while column 5 includes the two latter variables in the same model.

The results concerning the internal stock of knowledge help to confirm and qualify the Schumpeterian hypothesis. The stock of patents (*PStock*) of each firm exerts in fact a strong negative and significant effect on the costs of knowledge ($p < 0.01$ in all estimations). This is fully consistent with expectations, as the dependent variable is a measure of the unit costs of knowledge, which is likely to decrease as the stock of internal knowledge that firms can mobilize and use to generate new technological knowledge increases, other things being equal. Knowledge cumulability and non-exhaustibility exert a strong non-ergodic effect that favours incumbents that can rely on their internal knowledge in the recombinant knowledge generation process. We interpret these results in light of the Schumpeterian hypothesis. The characteristics of knowledge, combined with the recombinant feature of the knowledge generation process, favour incumbents that can prolong through time the benefits of earlier 'inventions'.

The intensity of R&D expenses is positively and significantly related to the cost of knowledge in all the estimations. This is quite in line with the expectations, R&D intensity being a measure of the technological efforts of the firm. The size of the estimated parameter however seems most important. It is consistently much lower than 1, ranging between 0.190 and 0.211 according to the different specifications. This suggests that the unit costs of knowledge increase, albeit much less than proportionately, with the levels of R&D intensity. The results of Table 6.10 (see below for more detail) where the baseline model is implemented with the absolute levels of R&D expenses and the explicit integration of the size of firms, provide further confirmation of our new interpretation of the Schumpeterian hypothesis: while the size of the stock of internal knowledge helps reduce the unit cost of knowledge, the amount of R&D expenditures and the absolute size of firms in terms of employment have a positive impact on the unit cost of knowledge. For a given size of the stock of internal knowledge larger firms have higher knowledge unit costs than smaller firms.

The results of the other variables are most important as they confirm the Marshallian hypothesis that knowledge costs decline with the size of the local pools of external knowledge. Moreover, they confirm that not only the size of the knowledge base but also its nature is of some significance.

The results of the variables that account for the size and composition of the regional knowledge base differ whether they concern the size of the external stock, measured using the stock of patents of the firms localized in the region, or the knowledge structure in terms of variety (*RegTV*), complementarity (*RegCD*) and similarity (*RegCOH*). If we focus on column 2 of Table 6.5, results show that the size of the regional knowledge stock (*RegPStock*) exerts a negative and significant effect on the cost of knowledge. This would suggest that companies that can access large pools of external knowledge save on the costs of their internal knowledge generating activities. As far as knowledge variety is concerned, results show that *RegTV* is positively and significantly related to the firm's cost of knowledge. Let us recall that this index provides a measure of the diversification of observed combinations of technologies in regions' knowledge bases. The results thus indicate that the higher the variety in the combination of technologies in the firm's region, the higher the cost associated with a firm's knowledge output. This might be due to the fact that firms need to put more effort into trying and experimenting with new combinations of technologies distributed across a wide range of technology domains.

When we disentangle the effects of related and unrelated variety we find that only the latter (*RegUTV*) is significant (as shown in columns 3 to 5). The procedure by which the index is derived (see Antonelli and Colombelli, 2015a, b for methodological details) reveals that the concepts of 'related' and 'unrelated' variety refer basically to the belonging of technologies to the same technological domain, as defined by the classification system used (in our case the IPC). The positive and significant impact of *RegUTV* on the cost of knowledge would imply that an increase in the regional variety of technologies that belong to very different technological domains is likely to increase the costs of knowledge generating activities at the firm level. The unit cost of knowledge increases as an effect of the higher volume of resources that the firm needs to commit in order to search for and absorb the locally available external knowledge.

The evidence concerning the effect of regional cognitive distance confirms such results. The coefficient is indeed positive and significant across all of the four models in which it is included. The cognitive distance may be interpreted as an index of the average dissimilarity among the different technological competences that make up the regional knowledge base. When the local knowledge that firms may access features technologies which are far away from one another in the technological space, firms need to strengthen their absorptive capacity by widening the scope of technological domains that they can master in order to take advantage of knowledge spillovers in the generation of new knowledge. This implies increasing volumes of firm-level R&D expenditures per single patent.

As a robustness check, we further estimated an extended model including the patenting activities of firms localized outside their region (*WRegPStock*). Table 6.6 reports the results of the fixed effects regression estimations for the equations including *WRegPStock*. These results confirm the robustness of our analysis as regards the variables included in the baseline model. Yet, *WRegPStock* turns out not to be significantly related to the cost of knowledge.

To further check the robustness of our analysis and to explore the different facets of the hypothesis, we ran additional models. Table 6.7 shows results for the equation excluding *RegPStock* from the covariates. Indeed, one may notice that such a variable has high correlation with regional

Table 6.6   Results: extended model

Variables	(1) PCost	(2) PCost	(3) PCost	(4) PCost	(5) PCost
R&D	0.211***	0.192**	0.201***	0.186**	0.183**
	(0.0763)	(0.0772)	(0.0771)	(0.0774)	(0.0775)
PStock	−0.489***	−0.459***	−0.458***	−0.465***	−0.464***
	(0.0605)	(0.0616)	(0.0617)	(0.0616)	(0.0616)
Age	0.0145	0.0904	0.0711	0.0910	0.103
	(0.160)	(0.163)	(0.163)	(0.163)	(0.163)
RegPStock		−0.582	−0.494	−0.288	−0.440
		(0.381)	(0.386)	(0.358)	(0.386)
WRegPStock		−1.513	−1.918	−2.260	−1.736
		(2.259)	(2.262)	(2.203)	(2.259)
RegTV		2.332**			
		(1.184)			
RegRTV			1.126		0.990
			(0.944)		(0.945)
RegUTV				1.785**	1.718**
				(0.868)	(0.870)
RegCD		8.297*	7.872*	8.347*	8.545**
		(4.257)	(4.256)	(4.257)	(4.261)
RegCOH		0.258	0.180	0.0782	0.194
		(0.239)	(0.242)	(0.214)	(0.241)
Constant	6.951***	20.65	26.12	27.57	22.90
	(0.881)	(17.93)	(17.69)	(17.16)	(17.72)
Observations	870	870	870	870	870
R-squared	0.386	0.395	0.393	0.396	0.397
Number of id	171	171	171	171	171

Notes:
Standard errors in parentheses.
*** $p < 0.01$, ** $p < 0.05$, * $p < 0.1$.

total, related and unrelated variety. This may affect the significance level of *RegPStock*. For this reason, we also ran the regressions by dropping regional knowledge stock, so as to check the robustness of the results concerning the variety measures. The results actually do not change.

In order to control for the relatively high correlation between *R&D* and *PStock*, Table 6.8 reports results for the regression models which exclude R&D from the vector of covariates. The results confirm the robustness of our model.

Finally, as a further robustness check and to better test the Schumpeterian

Table 6.7 Alternative specifications

Variables	(1) PCost	(2) PCost	(3) PCost	(4) PCost	(5) PCost
R&D	0.211***	0.197**	0.207***	0.193**	0.188**
	(0.0763)	(0.0769)	(0.0767)	(0.0768)	(0.0770)
PStock	−0.489***	−0.475***	−0.473***	−0.478***	−0.478***
	(0.0605)	(0.0610)	(0.0611)	(0.0610)	(0.0610)
Age	0.0145	0.0518	0.0411	0.0777	0.0786
	(0.160)	(0.161)	(0.161)	(0.161)	(0.161)
RegTV		1.897*			
		(1.094)			
RegRTV			0.841		0.731
			(0.865)		(0.864)
RegUTV				1.881**	1.837**
				(0.865)	(0.867)
RegCD		7.752*	7.679*	8.633**	8.431**
		(4.104)	(4.116)	(4.113)	(4.120)
RegCOH		0.182	0.115	0.0603	0.139
		(0.231)	(0.233)	(0.213)	(0.232)
Constant	6.951***	4.470	7.192***	7.243***	5.733**
	(0.881)	(2.974)	(2.337)	(1.651)	(2.431)
Observations	870	870	870	870	870
R-squared	0.386	0.393	0.391	0.394	0.395
Number of id	171	171	171	171	171

Notes:
Standard errors in parentheses.
*** $p < 0.01$, ** $p < 0.05$, * $p < 0.1$.

hypothesis about the positive effects of the size of firms, we also estimated additional models including firm *Size* among the covariates. Table 6.9 shows results for an extended version of our baseline model which, together with R&D intensity, includes *Size* as an independent variable. Here, again, the estimated parameter of R&D intensity, now taking into account the inclusion of the variable *Size*, is positive, albeit well below 1. The results in Table 6.10 provide the definitive test of our hypothesis and conclude the investigation on the effects of company size on the cost of knowledge. The results for this alternative specification – which includes *Size* and where R&D expenditures and patent stocks are measured in absolute terms instead of as a ratio of total assets – confirm that even when the size of the firm (in terms of employment) is directly included in the estimation model, the value of the estimated parameter of R&D

Table 6.8   Alternative specifications

Fixed effects	(1)	(2)	(3)	(4)	(5)
Variables	*PCost*	*PCost*	*PCost*	*PCost*	*PCost*
PStock	−0.446***	−0.424***	−0.421***	−0.433***	−0.431***
	(0.0587)	(0.0596)	(0.0598)	(0.0598)	(0.0598)
Age	−0.00375	0.0818	0.0598	0.0841	0.0978
	(0.161)	(0.163)	(0.163)	(0.163)	(0.164)
RegPStock		−0.632*	−0.548	−0.292	−0.479
		(0.376)	(0.381)	(0.357)	(0.380)
RegTV		2.846**			
		(1.149)			
RegRTV			1.487		1.291
			(0.922)		(0.923)
RegUTV				2.109**	1.997**
				(0.863)	(0.866)
RegCD		8.573**	8.226**	8.969**	8.965**
		(4.171)	(4.179)	(4.181)	(4.178)
RegCOH		0.340	0.261	0.120	0.269
		(0.237)	(0.240)	(0.215)	(0.240)
Constant	6.774***	7.877**	10.67***	9.495**	8.625**
	(0.883)	(3.981)	(3.728)	(3.772)	(3.820)
Observations	870	870	870	870	870
R-squared	0.380	0.389	0.386	0.389	0.391
Number of id	171	171	171	171	171

expenditures in absolute terms is positive, albeit well below 1. With respect to the results of the estimates of the baseline model conducted on R&D intensity, the increase is confirmed, although the estimated parameter now varies between 0.288 and 0.312. Also, *Size* is found to be positive related to the cost of knowledge.

The results of both the variables that account for the size of the firm and its R&D expenditures in absolute terms confirm our interpretation of the Schumpeterian hypothesis. Company size helps reduce knowledge unit costs only if it is measured in terms of the stock of internal knowledge. When size is measured in terms of R&D expenditures and employment, under the control of specific variables that account for the size of the internal knowledge stock, knowledge unit costs increase. The Schumpeterian hypothesis is not confirmed with respect to sheer size when knowledge generation costs are considered.

These results lead us to articulate the distinction between knowledge

Table 6.9   Alternative specifications

Variables	(1) PCost	(2) PCost	(3) PCost	(4) PCost	(5) PCost
R&D	0.211***	0.195**	0.207***	0.197**	0.189**
	(0.0794)	(0.0798)	(0.0796)	(0.0798)	(0.0800)
PStock	−0.476***	−0.444***	−0.444***	−0.454***	−0.449***
	(0.0647)	(0.0660)	(0.0661)	(0.0660)	(0.0660)
Age	−0.0991	0.00278	−0.0175	0.00831	0.0197
	(0.182)	(0.185)	(0.185)	(0.186)	(0.186)
Size	0.242**	0.250**	0.247**	0.238**	0.247**
	(0.117)	(0.117)	(0.118)	(0.117)	(0.117)
RegPStock		−0.743*	−0.669*	−0.423	−0.614
		(0.387)	(0.392)	(0.368)	(0.392)
RegTV		2.713**			
		(1.167)			
RegRTV			1.469		1.313
			(0.932)		(0.933)
RegUTV				1.868**	1.764**
				(0.878)	(0.880)
RegCD		8.274**	7.949*	8.661**	8.626**
		(4.211)	(4.218)	(4.225)	(4.222)
RegCOH		0.278	0.207	0.0708	0.223
		(0.244)	(0.246)	(0.221)	(0.246)
Constant	5.314***	7.648*	10.29***	9.439**	8.481**
	(1.206)	(4.149)	(3.889)	(3.928)	(3.984)
Observations	854	854	854	854	854
R-squared	0.386	0.395	0.393	0.395	0.396
Number of id	171	171	171	171	171

Notes:
Standard errors in parentheses.
*** $p < 0.01$, ** $p < 0.05$, * $p < 0.1$.

generation and knowledge exploitation (March, 1991). Further work might investigate the relationship between the sheer size of firms and knowledge exploitation. Our results suggest that for given levels of the internal stock of knowledge larger firms are less efficient in the generation of new knowledge. It becomes most important to understand whether larger firms might be more efficient in the exploitation of knowledge.

Table 6.10   Alternative specifications

Fixed effects	(1)	(2)	(3)	(4)	(5)
Variables	*PCost*	*PCost*	*PCost*	*PCost*	*PCost*
R&Dexp	0.312***	0.296***	0.309***	0.293***	0.288***
	(0.0780)	(0.0784)	(0.0781)	(0.0788)	(0.0788)
CumPatStock	−0.436***	−0.398***	−0.395***	−0.413***	−0.407***
	(0.0720)	(0.0737)	(0.0739)	(0.0740)	(0.0741)
Age	−0.0863	0.0222	0.00277	0.0267	0.0377
	(0.183)	(0.186)	(0.186)	(0.187)	(0.187)
Size	0.315**	0.312**	0.302**	0.307**	0.316**
	(0.123)	(0.124)	(0.124)	(0.124)	(0.124)
RegPStock		−0.812**	−0.748*	−0.505	−0.687*
		(0.391)	(0.396)	(0.372)	(0.397)
RegTV		2.541**			
		(1.173)			
RegRTV			1.382		1.233
			(0.936)		(0.938)
RegUTV				1.723*	1.621*
				(0.889)	(0.892)
RegCD		9.419**	9.101**	9.758**	9.744**
		(4.211)	(4.216)	(4.225)	(4.223)
RegCOH		0.258	0.194	0.0614	0.205
		(0.245)	(0.248)	(0.222)	(0.247)
Constant	7.204***	10.67***	13.05***	12.42***	11.57***
	(1.193)	(4.001)	(3.748)	(3.776)	(3.829)
Observations	854	854	854	854	854
R-squared	0.380	0.389	0.387	0.389	0.390
Number of id	171	171	171	171	171

Notes:
Standard errors in parentheses.
*** $p < 0.01$, ** $p < 0.05$, * $p < 0.1$.

4.   CONCLUSIONS AND IMPLICATIONS FOR FURTHER RESEARCH

The economics of knowledge has made major progress with the identification of the knowledge generation function. This empirical evidence has shown that the relationship between inputs and outputs of the innovative activity across firms exhibits huge variance. With given levels of R&D inputs, the actual amount of knowledge generated by each firm differs widely. A second important step along this line of analysis can be taken by analysing the knowledge cost function. This approach can help in understanding why the cost of knowledge is far from homogeneous. This evidence has been rarely detected in the literature and poorly investigated.

Study of the knowledge cost function enables us to analyse the role of the different cost items that concur with the definition of the knowledge output. This innovative approach enables us to explore from a novel perspective two important hypotheses that are at the core of the economics of knowledge: i) the so-called Schumpeterian hypothesis according to which firms with larger stocks of internal knowledge are superior in the generation of new knowledge; and ii) the so-called Marshallian hypothesis according to which knowledge externalities exert positive effects according not only to the density of the local pools of knowledge but also to their levels of coherence.

The empirical analysis of the costs of knowledge, based upon a panel of companies listed on UK and the main continental European financial markets (Germany, France, Italy and the Netherlands) for the period 1995–2006, for which information about patents has been gathered, has considered the unit costs of patents on the right hand side, and on the left hand side next to R&D expenditures, the stock of internal knowledge as well as the stock and the composition of external knowledge. The results confirm that the size and composition of the stock of internal knowledge play a key role in assessing the actual capability of each firm to generate new technological knowledge and hence in reducing the costs of knowledge. These results are important as they cast new light on the Schumpeterian hypothesis. Company size exerts positive – reducing – effects on knowledge unit costs only when it is measured by the stock of knowledge. The sheer size of firms in terms of R&D expenditures and employment, under the control of the size of the internal stock of knowledge, exerts positive – increasing – effects on knowledge unit costs. For given levels of internal knowledge, larger firms have higher knowledge unit costs than small firms. The sheer size of firms does not help in reducing knowledge cost generation. This result pushes us to reformulate the Schumpeterian hypothesis, introducing the distinction between knowledge generation and exploitation. We have demonstrated that sheer size does not help increase the cost-efficiency of firms: additional work should be done to investigate whether the sheer size of firms might favour the exploitation of knowledge. The results on the role of the size and composition of external knowledge fully confirm the Marshallian hypothesis, stressing the important role of the composition of the local knowledge pools.

These results bear important implications for technology policy at the regional level as well as for the strategic management of firms. Technology policy represents indeed one of the key levers that policymakers may use to trigger local development. Due to the collective and systemic nature of knowledge generation activities, the choice of the correct policy mix is crucial. The promotion of specific technological domains at the local level may impact the effectiveness of knowledge generation processes of incumbent firms. In this way, attempts to foster the emergence of technologies which are not consistent with the competences accumulated in the region are likely to increase the average level of unrelated variety and dissimilarity and, as a consequence, increase the average cost per patent. The implementation of technology policies that focus local knowledge endowments and try to upgrade them may be more effective than the pursuit of technological goals that are unrelated to the local pools of competence.

From the managerial viewpoint, the results of our analysis confirm the intuition of Edith Penrose about the central role of the stock of knowledge internal to each firm. The results also confirm that the composition of the bundle of technological activities carried out at the local level plays an important role. This has two important implications for decision-making. First, footloose firms should take into account, in their decisions concerning the location of their R&D laboratories, the local mix of technological competences so as to select sites with the size and mix of local knowledge pools that are more consistent with their technological strategies. Second, firms with a given rooted location should choose, among different possible technological strategies, those that are more compatible and consistent with the specific composition of the local knowledge pool. The location in areas featuring a bundle of technological competencies consistent with the innovation strategies of the firm is indeed likely to make the search process for new combinations of technologies more effective, and hence new knowledge less costly.

NOTES

*   The authors acknowledge the useful comments of the editor and two anonymous referees, as well as the financial support of the European Union DG. Research with the Grant number 266959 to the research project 'Policy Incentives for the Creation of Knowledge: Methods and Evidence' (PICK-ME), within the context Cooperation Program/Theme 8/Socio-economic Sciences and Humanities (SSH), and the institutional support of the Collegio Carlo Alberto.

1.   The implementation of the IPER database was financed by the Collegio Carlo Alberto under the IPER project.
2.   A 15 per cent obsolescence rate is the most common value used in the literature (see, for example, Nesta, 2008; Colombelli et al. 2013b). As a robustness check we also experimented with alternative obsolescence rates. We found that obsolescence rate makes little difference in empirical estimations.

7. The cost of knowledge and productivity growth

Cristiano Antonelli and Agnieszka Gehringer*

1. INTRODUCTION

Economic growth based on the continuous generation of innovation is a major contribution of the Schumpeterian legacy and endogenous growth models (Romer, 1990; Rivera-Batiz and Romer, 1991; Grossman and Helpman, 1991; Aghion and Howitt, 1992, 1998). The subsequent investigation of causes of productivity growth and of differences across regions and industries in the productivity dynamics attracted much attention in past empirical investigations. The intensive search for the determinants of total factor productivity (TFP) led to identifying such factors as R&D efforts, human capital accumulation, trade openness and financial globalization (Cameron et al., 2005; Biatour and Dumont, 2011; Gehringer, 2013; Gehringer et al., 2016b).[1]

Little attention has been paid to the huge differences in the costs of knowledge. Yet, the cost of knowledge differs widely over time and across countries that share similar regimes of intellectual property rights (IPR). We can assume, in fact, that the levels of knowledge appropriability are homogeneous across OECD countries. At the macro level, the average unit cost of knowledge was ranging between 0.2 million of constant purchasing power parity (PPP) $ in South Korea and New Zealand and 6.2 million in Belgium between 1985 and 2010. Moreover, such costs were varying over time in the majority of the OECD countries in our sample: it was generally lower in the last decades of the twentieth century and increased significantly afterwards. We integrate this stylized fact into the analysis of productivity dynamics and treat the cost of knowledge as a new determining factor of TFP growth. Accordingly, our main hypothesis is based on the following argumentation. Where the cost of knowledge is higher, more intensive R&D expenditures are needed to generate a piece of technological knowledge. Conversely, where it costs less to engage in innovative activities, relatively more new knowledge will be generated, contributing to faster productivity growth.

This hypothesis seems at first glance to be confirmed by the data, both for individual countries and for the average of the OECD sample. Regarding more precisely the development over time, although by their nature the TFP growth rates were more volatile, the steady increase in the unit cost of knowledge was accompanied by a diminishing trend of productivity growth.

The theoretical explanation of the possible link between the cost of knowledge and the productivity dynamics relies on the notion of pecuniary knowledge externalities, according to which external technological knowledge, regarded as indispensable in the generation of new technological knowledge and the eventual introduction of innovations, can be accessed and used at costs that are below equilibrium levels in some specific regional, industrial and institutional circumstances. When the private costs of knowledge lie below its social value, the unit cost of knowledge-intensive production is lower and the output larger than expected. The more intensive are knowledge externalities, the faster the productivity growth of the system at large.

Our approach and the main contribution to the literature are twofold. First, we construct a broad theoretical background showing that both the Schumpeterian and the Arrovian traditions of analysis support our working hypothesis. We build here upon two complementary models – the knowledge generation function (Nelson, 1982; Weitzman, 1996, 1998) and the material goods production function (Griliches, 1979) – that are put into a unified analytical framework. Subsequently, we empirically test our model in a panel investigation of 20 OECD countries between 1985 and 2010.

The rest of the chapter is organized as follows. In Section 2, we describe the relevant literature creating the conceptual context for our later analysis. Section 3 presents the theoretical model, which is then empirically tested in Section 4. Section 5 provides crucial policy implications, and the last section concludes.

2.   THE ANALYTICAL CONTEXT

Appreciation of the central role of knowledge externalities in the introduction of productivity-enhancing innovations is the result of the recent advances of two quite separate and yet complementary analytical approaches: i) the 'rediscovery' of the Schumpeterian notion of creative reaction contingent upon the availability of knowledge externalities; and ii) the Arrovian analysis of technological knowledge. Let us analyse them in turn.

2.1 The Schumpeterian 'Creative Reaction'

The new appreciation of the essay 'The creative response in economic history' published by Joseph Alois Schumpeter in the *Journal of Economic History* in 1947 and 'almost' forgotten since then enables the reappraisal of the Schumpeterian legacy. The Schumpeterian literature has paid much attention to his earlier contribution which introduced the notions of limited appropriability of knowledge and the transient duration of the monopolistic market powers associated with the introduction of innovations. According to Schumpeter (1942) imitative entry cannot be impeded, but only delayed. With respect to the extensive literature that impinged upon *Capitalism, Socialism and Democracy*, the attention paid to 'The creative response in economic history' has been inconsistent.

Yet Schumpeter (1947) provided quite an original framework where innovation is conceptualized as a creative response rather than the result of a routine or a rational plan. The availability of knowledge externalities is regarded as an indispensable condition to support the creative reaction and actually introduce technological innovations. According to Schumpeter, firms are often exposed to mismatches between the plans that are necessary to organize their current business and the actual conditions of product and factor markets. Their reaction can be adaptive (or passive) and creative. Passive reactions consist in textbook switching activities on the existing maps of isoquants and adjustment of prices to quantities, and vice versa. Passive reactions take place when firms cannot take advantage of knowledge externalities. Without pecuniary knowledge externalities, in fact, firms might be able to change their products and their processes, but they cannot introduce technological innovations. Without pecuniary knowledge externalities firms may increase the variety of their products and production processes, but cannot introduce productivity-enhancing innovations (Antonelli, 2008).

Their reactions to unexpected mismatches between expectations and actual conditions of both product and factor markets are creative when and if relevant pecuniary knowledge externalities are available. Only with pecuniary knowledge externalities can firms that try to react to mismatches that push them in out-of-equilibrium conditions introduce actual productivity-enhancing technological innovations (Antonelli, 2008, 2011).

Schumpeter's later essay uncovers the positive effects of the limited appropriability of knowledge: knowledge spills as it cannot be fully appropriated by the 'inventor' but yields positive externalities that can benefit all inventors. The integration of both the contributions of Schumpeter enables us to elaborate the distinction between positive and negative pecuniary knowledge externalities. Negative pecuniary knowledge externalities consist in the reduction of the revenues associated with the introduction of an innovation. Positive pecuniary knowledge externalities consist in the reduction of the costs of knowledge that are necessary to introduce an innovation. The algebraic sum defines whether the net pecuniary knowledge externalities are positive or negative.

Reappraisal of the Schumpeterian literature based upon appreciation of the notion of innovation as a creative response contingent upon the availability of net pecuniary knowledge externalities complements and integrates the new developments of the Arrovian economics of knowledge.

2.2   The New Economics of Knowledge

Technological knowledge is a peculiar good as well as a highly idiosyncratic activity. Following the path-breaking analyses by Richard Nelson and Kenneth Arrow it is well known that technological knowledge, as an economic good, is characterized by limited appropriability, non-exhaustivity, non-rival use, high levels of tacitness and substantial non-divisibility (Nelson, 1959; Arrow, 1962, 1969). Technological knowledge is at the same time the output of a dedicated activity and a necessary input not only in the production of other goods, but also in the generation of new technological knowledge (David, 1993; Crépon et al., 1998). The generation of technological knowledge, as a consequence, is shaped by the systematic recombination of the existing bits of knowledge together with current efforts of research, development and learning (R&D&L) activities (Weitzman, 1996, 1998; Lucas, 2008).

Intrinsic external and internal cumulability shapes the recombinant generation of technological knowledge. The basic inputs in the generation of technological knowledge are not only the current flows of R&D&L activities performed by each firm, but also the stock of existing knowledge generated internally by each firm and externally by the rest of the system (Saviotti, 2007).

Due to the very nature of technological knowledge, its owners are unable to fully appropriate their intellectual property and, consequently, knowledge spills to the rest of the system. Because of this limited appropriability, knowledge externalities take place (Griliches, 1979, 1992). Knowledge externalities, however, only rarely take the form of pure externalities. Indeed, access to external sources of knowledge is not entirely free: because of relevant absorption costs, the case of pecuniary knowledge externalities applies instead (Cohen and Levinthal, 1990).[2]

Absorption efforts are necessary to screen the existing knowledge of the other firms, identify the components that are indispensable to pursue the ongoing generation of new knowledge, access and retrieve them, and, finally, apply their content (Cohen and Levinthal, 1990). Because of the irreducible content of tacitness, access to the relevant portions of existing knowledge is possible only by means of systematic interactions and effective communication between knowledge producers and knowledge users. Knowledge interactions on their own are far from free: their successful implementation requires dedicated activities that entail specific costs (Lane et al., 2009). The cost of such interactions – involving screening, learning, communicating and absorbing – is, nevertheless, below the social value of knowledge exchanged, so that pecuniary knowledge externalities are possible (Gehringer, 2011a, b).

The portions of technological knowledge acquired externally are strictly complementary to the internal knowledge stock and to the current efforts of R&D&L in the recombinant generation of new technological knowledge. This means that firms cannot afford to be without external knowledge. Additionally, external knowledge brings about benefits in terms of pecuniary knowledge externalities. When and if pecuniary knowledge externalities are available and hence can be accessed, implying costs of external knowledge below the equilibrium levels, firms can introduce technological innovations that increase TFP (Antonelli, 2013a).

2.3 Pecuniary Knowledge Externalities and Productivity Growth

The two lines of enquiry converge with different motivations and in different analytical traditions – respectively the equilibrium and the out-of-equilibrium approach – to articulating the same hypothesis. When firms face unexpected product and factor market conditions, the costs of knowledge play a key role in shaping their reaction. When the costs of knowledge are high, firms cope with the out-of-equilibrium conditions by means of substitution processes: they try to adjust quantities to prices moving along the existing maps of isoquants. When the costs of knowledge are low, firms try to generate new technological knowledge so as to introduce innovations that change the existing map of isoquants. Their reaction can be creative only when and if the costs of knowledge are low.

Low knowledge costs make it possible to try to introduce innovations rather than adjusting quantities to prices, and vice versa. In order to introduce innovations, firms activate their knowledge generation processes and make effective use of the external knowledge that is available at costs that are below equilibrium levels. Pecuniary knowledge externalities are found in innovation systems characterized by high levels of knowledge connectivity when firms can access and use the existing stocks of knowledge at low costs. High knowledge connectivity in turn depends upon the quality of knowledge interactions not only between users and producers, but also at large between the various agents that are part of the system, including firms and research organizations. Strategic alliances aimed at implementing the knowledge capability of firms are key determinants of the general levels of knowledge connectivity.

Total factor productivity can be explained only in terms of access to pecuniary knowledge externalities. Pecuniary knowledge externalities in fact provide access to external knowledge at costs that are below equilibrium levels. This enables firms to cope with unexpected changes in product and factor market conditions by means of technological and organizational innovations.

Total factor productivity measures the mismatch between historic levels actually experienced and the levels of expected (equilibrium) output. Pecuniary knowledge externalities can account for the residual. When the access and use of the external stock of knowledge cost less than in the market equilibrium conditions, the reaction of firms can be creative and the actual output levels are larger than the expected ones. Here the distinction between pure or technical externalities and pecuniary knowledge externalities plays a central role. The stock of existing knowledge can be accessed and used as an indispensable input in the generation of further knowledge only with the intentional effort of prospective users. The access itself does not fall from heaven. It is the result of intentional and dedicated activities that entail costs. Such costs in turn depend not only on the sheer size of the stock of knowledge but also on the structural characteristics of the system and on the quality of knowledge-governance mechanisms that are in place. If we retain the notion of economic systems as rugged landscapes articulated by Krugman (1994, 1995) we see that some systems are endowed with landscapes that are better able than others to make the access to existing knowledge easier and less expensive, with larger knowledge externalities that in turn yield larger occurrence of creative reactions, decisions to innovate and, ultimately, higher rates of TFP increase.

3.   THE COST OF KNOWLEDGE AND PRODUCTIVITY GROWTH

A simple model whereby pecuniary knowledge externalities explain both TFP levels and increase can be stylized by means of two nested activities: a knowledge generation function and a technology production function. Both activities are usually performed by each and the same firm.

Our knowledge generation function follows analyses by Nelson (1982) and Weitzman (1996, 1998). The stocks of all existing technological knowledge both internal and external to each firm are indispensable, strictly complementary inputs, together with R&D activities and the valorization of learning processes, in the recombinant generation of new technological knowledge.

Internal and external stocks of existing knowledge can be accessed only if and when dedicated resources have been expended. We can thus write the recombinant knowledge generation function and the cost equation of technological knowledge of each firm. The knowledge generation function is a standard Cobb–Douglas and takes the following form:

$$TK(ISK, ESK, R\&D\&L) = ISK^A ESK^B R\&D\&L^\Delta \qquad (7.1)$$

where TK represents new technological knowledge generated with constant returns to scale by means of the current efforts in research, development and learning ($R\&D\&L$), the internal stock of knowledge (ISK) and the external stock of knowledge (ESK) – that is, the three indispensable productive factors. A, B and Δ are the respective output elasticities of the three inputs. The marginal rates of technical substitution of the three factors are:

$$\frac{\partial TK/\partial ISK}{\partial TK/\partial ESK} = \frac{A \cdot ISK^{A-1} ESK^B R\&D\&L^\Delta}{B \cdot ISK^A ESK^{B-1} R\&D\&L^\Delta} = (A/B)(ESK/ISK)$$

$$\frac{\partial TK/\partial ISK}{\partial TK/\partial R\&D\&L} = \frac{A \cdot ISK^{A-1} ESK^B R\&D\&L^\Delta}{\Delta \cdot ISK^A ESK^B R\&D\&L^{\Delta-1}} = (A/\Delta)(R\&D\&L/ISK)$$
$$(7.2)$$

$$\frac{\partial TK/\partial R\&D\&L}{\partial TK/\partial ESK} = \frac{\Delta \cdot ISK^A ESK^B R\&D\&L^{\Delta-1}}{B \cdot ISK^A ESK^{B-1} R\&D\&L^\Delta} = (\Delta/B)(ESK/R\&D\&L)$$

Denoting with t, u and z the unit price of the respective factors, the cost equation is:

$$C = tISK + uESK + zR\&D\&L \qquad (7.3)$$

Profits can be defined as:

$$\pi(TK) = sTK(ISK, ESK, R\&D\&L) - tISK - uESK - zR\&D\&L \qquad (7.4)$$

where s is the price of the knowledge output.

The first order conditions of profit maximization can be obtained by deriving equation (7.4) with respect to ISK, ESK and $R\&D\&L$, and putting the partial derivatives equal to zero. This can be expressed as:

$$\partial\pi/\partial ISK = s \cdot A \cdot ISK^{A-1}ESK^{B}R\&D\&L^{\Delta} - t = 0$$

$$\partial\pi/\partial ESK = s \cdot B \cdot ISK^{A}ESK^{B-1}R\&D\&L^{\Delta} - u = 0 \qquad (7.5)$$

$$\partial\pi/\partial R\&D\&L = s \cdot \Delta \cdot ISK^{A}ESK^{B}R\&D\&L^{\Delta-1} - z = 0.$$

From equation (7.5), the equilibrium conditions can be rewritten as:

$$t/u = (A/B)(ESK/ISK)$$

$$z/t = (A/\Delta)(R\&D\&L/ISK) \qquad (7.6)$$

$$z/u = (\Delta/B)(ESK/R\&D\&L)$$

It follows that firms select the equilibrium mix of inputs such that the relative unit costs are equal to the marginal rate of technical substitution. In some localized, historic, institutional and spatial circumstances, because of the quality of knowledge governance mechanisms at work and the high quality of knowledge interactions and communication, the unit costs of the access and reuse of the stock of existing knowledge are very low: hence the reaction of firms can be creative. Pecuniary knowledge externalities are found where and when the localized costs of the stock of external knowledge (u) are below socially desirable equilibrium – average – levels (u^*).

According to equation (7.6), it is in fact clear that when pecuniary knowledge externalities apply, given a certain budget, the firm will choose a combination of inputs biased towards a larger stock of external knowledge and a smaller internal stock of knowledge and of current efforts in R&D. When pecuniary knowledge externalities apply, as a consequence, the firm will be able to generate an amount of technological knowledge that is larger than in equilibrium conditions, and will be able to react creatively so as to generate new technological knowledge at low costs and actually introduce technological innovations.

The effects of the larger amount of technological knowledge generated at costs that are lower than in equilibrium conditions has a direct effect on the downstream production of all other goods. The recombinant knowledge generation function feeds the technology production function.

Following Griliches (1979), in fact, technological knowledge enters directly a standard Cobb–Douglas production function of all the other goods with constant returns to scale of each firm. Hence:

$$Y(K,L,TK) = K^{\alpha}L^{\beta}TK^{\gamma} \qquad (7.7)$$

where K, L and TK are the productive factors and α, β and γ the respective output elasticities. The marginal rates of technical substitution are:

$$\frac{\partial Y/\partial L}{\partial Y/\partial K} = \frac{\beta \cdot K^\alpha L^{\beta-1} TK^\gamma}{\alpha \cdot K^{\alpha-1} L^\beta TK^\gamma} = (\beta/\alpha)(K/L)$$

$$\frac{\partial Y/\partial L}{\partial Y/\partial TK} = \frac{\beta \cdot K^\alpha L^{\beta-1} TK^\gamma}{\gamma \cdot K^\alpha L^\beta TK^{\gamma-1}} = (\beta/\gamma)(TK/L) \qquad (7.8)$$

$$\frac{\partial Y/\partial TK}{\partial Y/\partial K} = \frac{\gamma \cdot K^\alpha L^\beta TK^{\gamma-1}}{\alpha \cdot K^{\alpha-1} L^\beta TK^\gamma} = (\gamma/\alpha)(K/TK)$$

Denoting with r, w and s the unit price of the three indispensable production factors in the production of Y, the cost equation is:

$$C = rK + wL + sTK \qquad (7.9)$$

Profits can be defined as:

$$\pi(Y) = pY(K, L, TK) - rK - wL - sTK \qquad (7.10)$$

where p is the price of output Y.

The first-order conditions can be obtained by deriving equation (7.10) with respect to factors K, L and TK, and putting the partial derivatives equal to zero. This can be expressed as:

$$\partial\pi/\partial K = p \cdot \alpha \cdot K^{\alpha-1} L^\beta TK^\gamma - r = 0$$

$$\partial\pi/\partial L = p \cdot \beta \cdot K^\alpha L^{\beta-1} TK^\gamma - w = 0 \qquad (7.11)$$

$$\partial\pi/\partial TK = p \cdot \gamma \cdot K^\alpha L^\beta TK^{\gamma-1} - s = 0.$$

From equation (7.11), the equilibrium conditions are:

$$w/r = (\beta/\alpha)(K/L)$$

$$w/s = (\beta/\gamma)(TK/L) \qquad (7.12)$$

$$s/r = (\gamma/\alpha)(K/TK).$$

Firms select the equilibrium mix of inputs such that the relative unit costs are equal to the marginal rate of technical substitution. Corresponding to these three conditions, the profit-maximizing firm will identify TK^*, K^* and L^*, that is, the equilibrium levels of the production factors.

With positive pecuniary knowledge externalities in the upstream generation of technological knowledge and, hence, cheap localized costs of technological knowledge, below equilibrium level, $s < s^*$, firms will try to generate new technological knowledge in order to introduce innovations. Hence firms will use a technique characterized by higher levels of TK and lower levels of both capital and labour: they will use a more knowledge-intensive technique than the adaptive firms that cannot benefit from pecuniary knowledge externalities. Most importantly, the firms that benefit from pecuniary knowledge externalities will be able to react creatively to unexpected mismatches in both product and factor markets. This happens because they can generate new technological knowledge at a cost that is below equilibrium, and hence introduce productivity-enhancing technological innovations, experiencing an actual increase in their levels of TFP: producing an output Y that is larger (and cheaper) than in general equilibrium conditions.

Following Abramovitz (1956), we know that the level of TFP is measured by the ratio between the real historic levels of output Y and the theoretical ones calculated as the result of the equilibrium use of production factors:

$$A = Y/K^* L^* TK^* \qquad (7.13)$$

where K^*, L^* and TK^* are the general equilibrium quantities of production factors and A measures TFP.

Technological knowledge that has been generated without the availability of pecuniary knowledge externalities will yield equilibrium levels of output. In these conditions firms can introduce novelties rather than innovations. Novelties consist in an increase in the variety of products and processes, such as changes in production processes, higher levels of product differentiation with new characteristics of their products. Novelties differ from innovations. The former are produced in equilibrium conditions such that the marginal product of inputs matches their costs. Innovations instead yield TFP-enhancing effects (Link and Siegel, 2007).

The results of the modelling exercise can be summarized as follows: firms produce more than expected, and hence experience an 'unexplained' residual in the actual levels of output. The observed levels of output are larger than the expected ones ($Y > Y^*$) if and when the localized costs of the access and secondary use of the stock of existing technological knowledge in the upstream knowledge generation function are lower than in general equilibrium ($u < u^*$). It is clear in fact that, when pecuniary knowledge externalities apply, the output of the recombinant knowledge generation function in terms of technological knowledge is larger than in general equilibrium conditions and the costs of the technological knowledge that enters the downstream Cobb–Douglas technology production function for all the other goods are also lower ($s < s^*$). Moreover, the lower such costs, the stronger the effect of productivity increases. With a given budget, firms that benefit from pecuniary knowledge externalities are able to generate a larger amount of technological knowledge, and hence an amount of all the other goods that is larger than the expected levels based upon equilibrium assumptions.

In such conditions, qualified by positive pecuniary knowledge externalities, each firm operates in localized (and transient) equilibrium conditions, but the aggregate output of the system is larger than expected in general equilibrium conditions. The working of pecuniary knowledge externalities is compatible with short-term, instantaneous equilibrium conditions at the firm level, while at the aggregate level the system is far from equilibrium.

This approach enables us to take into account the specific characteristics of the rugged landscapes that make pecuniary knowledge externalities actually available. Pecuniary knowledge externalities, in fact, are not a ubiquitous, persistent and spontaneous attribute of any kind of economic system, at any time. Quite the opposite; pecuniary knowledge externalities are the endogenous result of the specific conditions of the rugged landscapes in which external knowledge flows and can be accessed. The actual access to external knowledge may take place in highly localized conditions when and where knowledge cumulability is actually implemented and supported. Such characteristics, in turn, are a consequence of the past generation of technological knowledge and introduction of technological innovations. As such they are endogenous to the system (Saviotti and Pyka, 2008).

4.   EMPIRICAL EVIDENCE

4.1   The Descriptive Evidence

The empirical analysis refers to a sample of 20 industrialized OECD countries that implement and enforce homogeneous intellectual property rights regimes, observed over the time span 1985–2010 (Table 7.1). We assume that the levels of knowledge appropriability are comparable across the 20 countries considered. The unit cost of knowledge, as measured by the ratio of R&D expenditure to patents, differs widely over time and across countries, regions, industries and firms. As Table 7.1 (column 1) shows, at the macro level, the unit cost of knowledge ranged from 0.2 million constant PPP $ in South Korea and New Zealand to 5.5 million in Portugal and 6.2 million in Belgium. The variance across the OECD average of 1.9 million is relevant. Moreover, such costs varied over time in the majority of countries in our sample: it was generally lower in the first half of the period and increased significantly afterwards.

This dynamic over time is clearly confirmed in Figure 7.1. The average

Table 7.1 *Unit cost of knowledge and TFP growth in OECD countries,*
              1985–2010

Country	Unit cost of knowledge			TFP growth (%)		
	1985–2010	1985–98	1999–2010	1985–2010	1985–98	1999–2010
Australia	0.5	0.5	0.6	0.9	1.1	0.5
Austria	1.8	1.1	2.6	1.4	1.3	1.4
Belgium	6.2	3.4	9.0	0.8	1.2	0.3
Canada	0.5	0.4	0.6	0.4	0.2	0.5
Denmark	1.7	1.1	2.5	0.8	1.1	0.3
Finland	1.5	0.6	2.6	1.8	2.2	1.2
France	2.1	1.9	2.4	1.0	1.4	0.6
Germany	1.1	1.1	1.1	0.9	1.2	0.7
Ireland	1.3	0.5	2.1	2.7	3.5	1.8
Italy	1.8	1.6	2.0	0.4	1.0	−0.3
Japan	0.3	0.3	0.3	1.4	1.8	0.9
Korea	0.2	0.3	0.2	3.8	4.3	3.3
Netherlands	3.4	2.9	3.9	1.0	0.9	1.1
New Zealand	0.2	0.1	0.2	0.5	0.5	0.4
Portugal	5.5	2.6	8.8	1.3	2.5	0.6
Spain	2.8	1.9	3.9	0.5	1.0	−0.1
Sweden	2.5	1.3	3.4	0.9	0.7	1.2
Switzerland	2.3	1.4	3.4	0.0	−0.6	0.4
UK	1.1	0.9	1.3	0.8	0.8	0.7
US	1.1	1.3	0.9	1.1	0.9	1.4
Average OECD	1.9	1.3	2.6	1.1	1.4	0.8

Note: Unit cost of knowledge is calculated as the ratio between the total R&D
expenditures in millions of constant (2005) PPP $ and the number of patent applications
made to the World Intellectual Property Organization (WIPO). TFP growth is the growth
rate of multi-factor productivity.

Source: Own calculations based on the IPER database built upon OECD STAN and
WIPO data.

unit cost of knowledge was steadily increasing between 1985 and 2010 for
our sample. For the growth of TFP the picture is less clear, although a
diminishing tendency over time can be recognized.

4.2   Estimation Strategy

Our focus is on the relationship between the unit cost of knowledge generation and TFP growth. This relationship might in principle go two ways. On
the one hand, and in line with our previous argument, where the cost of
knowledge is lower, the more intensive and stronger will be pecuniary knowledge externalities and the faster should be the growth of TFP. On the other

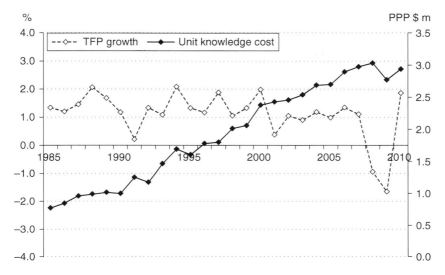

Note: See note of Table 7.1.

Source: Own calculations based on the IPER database built upon OECD STAN and WIPO data.

Figure 7.1 *Development of TFP growth and of the unit cost of knowledge on average of 20 OECD countries, 1985–2010*

hand, in a context characterized by a positive dynamics of TFP, the costs of knowledge generation are likely to decrease due to the more extensive availability of newly generated external knowledge. Although we believe that this second mechanism is less likely to be true, or at least not on a regular basis, we account for this potential simultaneity by applying the instrumental variable method (also called two-stage least squares, 2SLS), where the cost of knowledge is treated as an endogenous regressor and needs to be instrumented.

It is a common problem to find variables serving as valid instruments, in the sense of being satisfactorily exogenous with respect to the dependent variable and at the same time strongly correlated with the endogenous regressors. If instruments are weak, the precision of instrumental variable estimations is lower than that of simple OLS regressions. On the other hand, if endogenous regressors are included in the estimation, OLS results will be biased. As we suspect – and confirm with a test – that endogeneity could be an issue in our framework, we aim to explicitly deal with it, but at the same time ensure that our procedure does not lead to loss of precision of our estimations.

In order to implement this approach, our econometric model is based upon a system of two equations: the cost of knowledge equation and the TFP equation, where the second-step TFP growth equation includes the fitted values of knowledge cost from the first step. The two-step model can be represented in the following system:

$$CK_{it} = \alpha_1 + \alpha_2 STOCK_{it} + \alpha_3 + + \mathbf{V}_{it}'\boldsymbol{\alpha_4} + \alpha_5 CK_{it-1} + \alpha_6 CK_{it-2} + \tau_t + \rho_i + \varepsilon_{it}$$
$$(7.14)$$

$$\Delta TFP_{it} = \beta_1 + \beta_2 \widehat{CK}_{it} + \mathbf{V}_{it}'\boldsymbol{\beta_3} + \vartheta_t + \mu_i + \varepsilon_{it} \qquad (7.15)$$

The first step described in equation (7.14) consists in testing the key hypothesis that the cost of knowledge (*CK*) reflects the density of the stock of existing knowledge (*STOCK*). Each firm at each point in time will be better able to generate technological knowledge the larger the amount of pecuniary knowledge externalities. The latter in turn are likely to stem diachronically from the stock of existing knowledge. Hence we expect that the costs of knowledge in a country is lower, the larger is the stock of patents per capita. In addition to the two aforementioned instruments (the so-called excluded instruments), we include the first and second lag of the dependent variable. Vector $\mathbf{V}_{it}'$ is the set of exogenous covariates that should be included in both the first and the second stage (Angrist and Pischke, 2008). Finally, τ_t and ρ_i are the time- and country-specific effects, whereas ε_{it} is the indiosyncratic error term.

A comment is due here to our measure of the cost of knowledge. This is captured by the R&D expenditures made in a country per unit of patent, and thus most closely refers to what we called in our theoretical discussion 'internal cost of knowledge'. Ideally, we would be willing to have a measure of cost that is an average between internal and external cost of knowledge generation, where external cost refers to the cost necessary to support when acquiring knowledge from sources external to a firm's own business. However, there are insuperable limits in finding such a measure and we are constrained to sticking to R&D expenditures only.[3] This notwithstanding, even for our R&D unit cost variable, our main hypothesis always holds: given the availability of pecuniary knowledge externalities, in any instant of time firms face a cost advantage associated with external knowledge, which provides them with the incentive to adjust their production plans towards a higher relative implementation of external versus internal knowledge. It follows thus that the more abundant are knowledge externalities, the lower should be the unit cost of (internal) knowledge and the higher will be productivity growth.

In the second step, we account for the endogeneity of the costs of knowledge within the productivity equation. In order to validate our instrumenting strategy we implement different post-estimation diagnostics – precisely, weak instruments test, over-identification test, redundancy test (see Table 7.4). Additionally, we check whether our excluded instrumental variables affect the dependent variable only indirectly, by including them as regressors in the estimations of TFP growth.[4] We could confirm that these variables were not directly influential in TFP growth and, thus, the exclusion restriction cannot be rejected.[5]

More precisely regarding the second stage, in addition to our main explanatory variable, we account for other possible factors determining TFP growth, as suggested in the past literature. Finally, our econometric model assumes the form:

$$\Delta TFP_{it} = \beta_1 + \beta_2 \widehat{CK}_{it} + \beta_3 HC_{it} + \beta_4 TO_{it} + \beta_5 GDP/cap_{it} + \vartheta_t + \mu_i + \varepsilon_{it} \quad (7.16)$$

where ΔTFP_{it} is the annual growth rate of TFP in country i at time t; $\widehat{CK}_{it}$ is our variable of interest expressing the unit cost of knowledge generation; HC_{it} is a measure of human capital stock available in an economy; TO_{it} refers to a measure of trade openness; and GDP/cap_{it} refers to national income per capita. The variables and data sources are presented in Table 7.2, and detailed description of the method in Appendix B.

Table 7.2   Variables used in the estimations: description and data sources

Variable	Description	Source
$\Delta(TFP)$	Growth rate of TFP	OECD.stat
CK	Cost of knowledge, measured in terms of the ratio between R&D spending and the number of patent applications in a country each year in 1985–2010	Own calculation based on OECD.stat (R&D spending) and WIPO (patent applications)
TO	Openness to trade measured as the ratio between the sum of imports and exports over GDP	Penn World Table 7.1
HC	Human capital: R&D personnel per 1000 employees	OECD.stat
GDP/cap	GDP per capita	Own calculation based on Penn World Table 7.1
Excluded instrument		
STOCK	Patent stock per capita	Own calculations applying the perpetual inventory method on annual patent applications to WIPO*

Note: * A detailed description of the method is included in Appendix B.

Coefficient β_1 is a constant, whereas coefficients β_2 to β_5 are supposed to measure the marginal contribution of each factor to TFP growth. Finally, ϑ_t and μ_i are the time and country dummies, respectively, whereas ε_{it} is the idiosyncratic error term.

To justify our set of control variables, the motivation to include a measure of *trade openness*, as expressed in terms of the share of the overall trade volume (imports plus exports) relative to the size of the economy (GDP), derives from the fact that the intensifying trade integration between economies has often been argued to have a positive impact on productivity growth. This impact can be direct – when trade flows (particularly export) take place in technology-intensive sectors – and indirect – through the acquisition of specialized technological skills as a consequence of interactions between the trade partners (see López, 2005 for a survey of the related literature).

The inclusion of human capital directly derives from the models of endogenous growth, supportive for the positive role in the growth process played by the accumulation of skills (Lucas, 1998). We measure the stock of human capital as the ratio of the R&D personnel for 1000 employees. In that way, we aim to capture the influence of the skilled labour that is most intensively involved in innovative processes.[6] Finally, we include GDP per capita in order to account for the convergence process. More precisely, economies that are at a relatively lower stage of economic development (that is, lower GDP per capita) are supposed to experience relatively faster productivity growth. Alternatively, as another way to control for the possible differences in economic development and their influence on the speed of productivity growth, one could include the initial levels of TFP. Accordingly, we checked this hypothesis as well, but the coefficient on this variable, although reporting the right sign, was never significant.

Table 7.3 presents the summary statistics of the variables included in estimations. Regarding the main variables of interest, average annual growth rate of TFP was equal to 1.1 per cent, ranging between −7.8 per

Table 7.3   Summary statistics

Variable	Obs.	Mean	Std Dev.	Min.	Max.
$\Delta(TFP)$	469	1.11	1.71	−7.8	7.8
CK	451	1.96	2.08	0.09	10.93
TO	520	69.16	33.90	16.00	183.30
HC	404	10.77	4.05	2.44	24.73
GDP/cap	520	27,559.24	6731.22	7352.31	47,134.06
STOCK	484	7471.04	11,062.42	150.15	58,820.32
R&D	457	554.91	281.11	42.07	1265.10

cent and 7.8 per cent. The unit cost of knowledge was on average almost 2 million constant PPP $, with a minimum of 0.09 and a maximum of 10.9 million.

4.3 Results

Table 7.4 summarizes the results of the estimations of equation (7.16). We report the results from four methods. Column (1) shows the results from a pooled OLS method. In columns (2) and (3) we estimate analogous specifications with methods explicitly accounting for the panel structure of our dataset, namely random effects and fixed effects model. The results from these three methods serve to sort out the direction of the bias with respect to the instrumental variable estimates (2SLS) reported in columns (4)–(6).

More precisely regarding the IV estimations, we first estimate a model where we include only the instrumented knowledge cost variable (column 4). This is to show the direct impact of the variable of interest on TFP growth. In other words, this is the effect of the cost of knowledge disconnected from the influence coming from other factors. We then include such other factors in the subsequent two specifications. In column 5, openness to trade, human capital and GDP per capita are considered. Finally, in column 6, the latter variable is replaced by the initial value of TFP.

As the endogeneity test regarding the endogenous regressor (CK) in all cases and at all conventional significance levels clearly rejects the null hypothesis of exogeneity, the IV specifications are preferred over those run according to the alternative methods, OLS, RE and FE. Regarding the direction of bias, the latter methods seem to underestimate the effect of knowledge cost on TFP growth. Moreover, considering the other tests reported (Hansen J test of validity of the over-identifying restrictions and the Kleibergen–Paap test of weak instruments), they suggest that the IV estimations are correctly specified.[7]

The results clearly suggest that the effect of knowledge cost on TFP growth is negative, meaning that the higher the unit cost of generating new knowledge, the lower the rate of growth of TFP. In quantitative terms this impact ranges between −0.35 and −0.36, depending on the control variables included. If the cost of knowledge increases by one unit (corresponding to 1 million 2005 constant PPP $), this leads to a 0.35–0.36 percentage point decrease in the TFP growth rate. Consequently, given that between 1999 and 2010 the average unit cost of knowledge for the OECD countries in our sample increased by around 1.3 million PPP $, this contributed to a reduction in TFP growth rate of approximately 0.42 percentage points. If the cost of knowledge remained unchanged at the average level from

Table 7.4   Results of estimations of the link between cost of knowledge generation and productivity dynamics

	(1)	(2)	(3)	(4)	(5)	(6)
	POLS	RE	FE	IV	IV	IV
CK	-0.274***	-0.196***	-0.199***	-0.342***	-0.354***	-0.354***
	(0.044)	(0.071)	(0.087)	(0.090)	(0.079)	(0.079)
TO	0.018***	0.021***	0.029**		0.025***	0.025***
	(0.003)	(0.007)	(0.013)		(0.008)	(0.008)
HC	0.039*	0.062*	0.081		0.053	0.053
	(0.021)	(0.036)	(0.050)		(0.054)	(0.054)
GDP/cap	-0.120**	-0.089**	-0.088**		-0.144	
	(0.020)	(0.025)	(0.042)		(0.030)***	
Initial TFP						-51.05
						(82.64)
N. obs.	360	360	360	280	255	255
R-squared (overall)	0.422	0.389	0.337	0.585	0.632	0.632
Hansen J				0.111	0.993	0.993
Weak identification				185.64	145.18	142.00
Endogeneity test				0.014	0.009	0.009
Redundancy test				0.009	0.004	0.004
First-stage results for the excluded instruments						
STOCK				-0.00002***	-0.00002***	-0.00003***
				(5.73e-06)	(7.47e-06)	(7.47e-06)
CK_{t-1}				0.813***	0.788***	0.788***
				(0.050)	(0.053)	(0.053)
CK_{t-2}				-0.015	-0.067**	-0.067**
				(0.020)	(0.027)	(0.027)
Partial R-squared				0.848	0.837	0.837

Notes:

Dependent variable is TFP growth rate. In column 1, pooled OLS results are shown. In columns 2 and 3, panel data random effects (RE) and fixed effects (FE) models are estimated. In columns 4–6 IV (2SLS) estimations are reported.

The endogenous regressor, CK, is instrumented with its first and second lag, as well as with patent stock per capita and R&D expenditure per capita.

Weak identification test reports the Kleibergen–Paap rk Wald F statistics.

Hansen J is the Chi-squared p value of the over-identification test of all instruments.

Endogeneity test reports the p value of the Chi-squared test having under the null the hypothesis that CK is exogenous.

Redundancy test reports p value of the Chi-squared test that checks whether the excluded instruments (patent stock per capita and R&D expenditures per capita) are redundant.

Partial R-squared refers to the goodness of fit of the excluded instruments.

In all specifications, time- and country-specific effects are included. Heteroskedasticity and autocorrelation robust statistics are reported.

the pre-period (1985–98) then, *ceteris paribus*, TFP would have grown on average between 1999 and 2010 at the rate of 1.2 per cent.

Regarding the other factors explaining the TFP growth, trade openness was positively contributing to speeding it up. Human capital variable, although reporting the expected positive sign, remained always insignificant. Finally, the convergence hypothesis seems to be confirmed by the negative sign of the coefficient corresponding to GDP per capita. Accordingly, countries experiencing lower levels of GDP per capita could register higher TFP growth rates. This same direction of influence, although with insignificant estimated coefficient, is found for the variable measuring the initial level of TFP.

Finally, the results of the first-stage estimation relative to the excluded instruments are also reported. They confirm that the cost of knowledge diminishes as the stock of existing knowledge increases.

5.   POLICY IMPLICATIONS

The stock of existing knowledge is a primary factor of the levels of pecuniary knowledge externalities. In turn, pecuniary knowledge externalities are the cause of TFP growth. Access to the stock of existing knowledge, however, is neither free nor automatic nor homogeneous across economic systems. It depends upon the quality of the rugged landscapes in which the generation of technological knowledge takes place and it requires a wide range of dedicated activities, including searching, screening, decodifying, interacting and learning. The creative capacity of a system is influenced by its transfer and absorptive capacities. Here policy action can play an important role in implementing the appropriate mix of interventions so as to support the creative capacity of the system (Furman et al., 2002; Cincera et al., 2014).

Knowledge transfer policy interventions finalized to favour the transfer, dissemination and actual access to the stock of existing knowledge may be very effective as their ultimate effects are reduction in the cost of knowledge, increase in pecuniary knowledge externalities and the eventual increase in total factor productivity.

Such knowledge transfer policy interventions should primarily focus on the large public research infrastructures, with the aim of making the technological knowledge generated by the public research system more accessible to prospective users. The systematic encouragement of academic departments to enter the markets for knowledge-intensive business services (KIBS) can help small and medium-sized firms take advantage of the stock of knowledge existing within the academic walls. At the individual level, the extension of professors' privilege so as to favour their professional activity in the markets for KIBS should increase the interactions of scientific personnel with the business community and put in motion the exchange of creative and praxis-oriented ideas.

More intense interactions and transactions between the public research system and the business community may also help to better focus the research activity of the public research system, directing it towards scientific fields that are more likely to yield 'useful' knowledge. The identification of the appropriate portfolio of research activities able to pursue scientific progress and yet contribute to economic growth is a major problem of the public research system. The feedback from the business community expressed by the intensity of academic outsourcing and knowledge transactions may provide useful signals to grasp the actual scope of application of the different scientific fields.

With respect to private sector knowledge transfer, policy interventions should focus on intellectual property rights regimes. The introduction of subsidies for the purchase of IPR, together with classical subsidies for funding R&D, might help the dissemination of the existing stock of knowledge and its easier access, helping the deepening of transactions in the markets for IPR with positive effects on both the demand and the supply side. The further reduction of the constraints of competition policies can favour the cooperation between knowledge users and knowledge producers along value chains increasing knowledge interactions. As a third step, the structure of IPR regimes might be reconsidered with respect to their exclusivity. The IPR regimes play an important role. Exclusive, long-lasting IPR regimes may delay access to the stock of existing knowledge to the point that prospective users may prefer to reinvent existing but non-accessible knowledge. For given levels of the stock of existing knowledge the systems endowed with an organized architecture of knowledge interactions and transactions may experience higher levels of pecuniary knowledge externalities that systems where knowledge interactions and transactions are made expensive by lack of trust and excess opportunism.

According to a large body of literature, the weakening of intellectual property rights might entail a reduction in incentives to generate technological knowledge, with the ultimate perverse effect of increasing access to existing knowledge and hence reducing the costs of knowledge but decreasing the incentives and hence the amount of knowledge actually generated. A reduction of the exclusivity of IPR accompanied by the identification of correct levels of royalties, however, might help the dissemination of existing knowledge yet defend appropriability. Identification of the correct levels of royalties is clearly crucial to combine the need to secure the rewards to innovators with the goal of increasing as much as possible the social surplus stemming from the introduction of innovations (Antonelli, 2013b).

6.   CONCLUSION

Technological knowledge is a peculiar economic good and it is the output of a highly idiosyncratic economic activity. Technological knowledge is at the same time an output and an input, not only in the production of all other goods but also in the generation of new technological knowledge. Because of such peculiar characteristics of technological knowledge as an economic good – namely, intrinsic non-exhaustivity, cumulability and complementarity – the access and use of the entire stock of existing technological knowledge, both internal and external to each firm, is strictly necessary. The stock of existing technological knowledge is a necessary input strictly complementary to current research and development and learning activities for the generation of new technological knowledge.

The access conditions to the stock of existing knowledge play a crucial role in assessing the actual capability of firms to innovate. Knowledge cumulability applies both to the internal stocks of knowledge generated by each firm and to the external stocks of knowledge available in the economic system at large. No firm can generate new technological knowledge and introduce technological innovations starting from scratch. Nor can a firm command all the existing knowledge. The access and use of the stock of external knowledge – that is, the components of the total stock of knowledge that have been generated and are possessed by the other firms in the system – are non-disposable inputs, for nobody can command all the knowledge available at any point in time.

Economic systems where the access conditions to existing technological knowledge are better enjoy substantial pecuniary knowledge externalities: external knowledge in fact is accessed at costs that are below equilibrium levels. In such conditions, economic systems where the stock of existing knowledge both internal and external to each firm can be accessed in more cost-efficient conditions are able to react more creatively to unexpected out-of-equilibrium conditions. Firms succeed in coping with unexpected changes in product and factor markets by introducing technological innovations. As a consequence, it is clear that the costs of knowledge are far from homogeneous across countries and historic times. Countries that have been able to elaborate efficient knowledge-governance mechanisms, making it possible for firms to access and (re)use the existing stocks of knowledge, experience faster rates of introduction of technological innovations and higher levels of total factor productivity growth.

The empirical evidence, based upon a representative group of 20 OECD countries for the considerable stretch of 25 years, confirms that knowledge costs, as measured by the ratio of total R&D expenditure to patents, vary widely across countries and over time. Moreover, this variation explains much of the dynamics in TFP growth. In particular, where and when knowledge costs are lower, TFP growth is faster.

Pecuniary knowledge externalities are indeed external to each firm, but endogenous to the system. At each point in time and space, in fact, the levels of pecuniary knowledge externalities do not fall from heaven like manna; they depend upon the amount of innovative effort made by agents in the system and the structure of knowledge interactions and transactions that relate each firm to the others. The effects of past actions on current levels of pecuniary knowledge externalities may be both negative – due to congestion – and positive because of knowledge cumulability. When the effects are positive a self-feeding process of growth based upon positive feedback is likely to take place and persist until pecuniary knowledge externalities keep increasing.

The policy implications of our analysis are most important. The improvement of the knowledge-governance mechanisms should be a crucial ingredient of innovation policymaking. Such knowledge-governance mechanisms contribute to the reduction of the cost of knowledge generation and of the persistent access to the external stock of existing knowledge at low costs. Consequently, they yield substantial support to the eventual increase in the rate of introduction of innovations, and are able to account for the persistent increase in the general efficiency of the economic system.

NOTES

*   The authors acknowledge the institutional support of the research project 'Incentive Policies for European Research' (IPER) and the funding of the European Union DG Research with Grant number 266959 to the research project 'Policy Incentives for the Creation of Knowledge: Methods and Evidence' (PICK-ME), within the context of the Cooperation Program/Theme 8/Socio-economic Sciences and Humanities (SSH) at the Collegio Carlo Alberto and the University of Torino. This chapter benefited greatly from the insightful comments and suggestions of Gary Jefferson, Filippo Belloc and other participants at the INFER Annual Conference in Pescara, May 2014 and at the Governance of a Complex World Conference in Turin, June 2014.
1. See Gehringer et al. (2016a) for an overview of the relevant literature.
2. The distinction between pure (or technological) externalities and pecuniary externalities dates back to Tibor Scitovsky (1954). Pure externalities are found when inputs can be used at no costs. Pecuniary externalities take place when the actual costs of an input, in a specific and localized condition, are lower than in equilibrium conditions (see also Antonelli, 2008).
3. There are other problems with this measure of knowledge cost as well. First, not all internal R&D activities are aimed at patent production, as they might end in other

productive outcomes like trade secrets, trademarks or other legally unprotected forms of innovation. This potentially leads us to overestimate the true cost of knowledge. Second, not all patents needs formal R&D expenditure, yet other forms of innovative investment. There might be patented incremental innovations obtained without R&D. This could lead us to underestimate the real cost of knowledge. Finally, and related to the previous two points, we standardize our total cost of knowledge by the number of patents, whereas other non-patented innovations are common.

4.   For brevity, we do not report the results of the estimations here. They are available upon request.

5.   This is also confirmed by the very low pair-wise correlation between the excluded instruments and ΔTFP, as shown in the correlation matrix in Table 7A.1.

6.   It is not easy to find a good variable measuring the stock of human capital available in the economy. It has been common practice in the literature to measure human capital in terms of secondary school attainment. Nevertheless, it can be argued that this is a too broad definition and that more qualified and more specific skills are essential when trying to grasp the influence of human capital on productivity. Consequently, our measure of human capital expresses an attempt in this direction. However, it can be argued that human capital is a measure of R&D intensity and, as such, could be correlated with our measure of unit cost of knowledge. We checked for this possibility and found a very low and insignificant correlation between the two.

7.   More precisely, the Hansen J test regresses the residuals from the IV estimation on all instruments (included and excluded). The null hypothesis is then that all instruments are uncorrelated with the residuals. Thus, given that we cannot reject the null, we can conclude that our over-identifying restrictions are valid. The Kleibergen–Paap rk Wald F test statistic refers to the weak identification test proposed by Stock and Yogo (2005). The test statistic is confronted with (different levels of) the rejection rate r that the researcher tolerates. Under the null, the test rejects too often. To reject the null, the F test statistic must exceed the critical value. In our case, we are always able to reject the null.

APPENDIX A

Table 7A.1   Correlation matrix between excluded instruments and dependent variables

	ΔTFP	STOCK	CK	CK_{t-1}	CK_{t-2}
ΔTFP	1				
STOCK	−0.011	1			
CK	−0.193	−0.304	1		
CK_{t-1}	−0.190	−0.304	0.980	1	
CK_{t-2}	0.083	−0.069	0.052		

APPENDIX B

The calculation of the stock of patents is based on the perpetual inventory method, analogously linked in the case of calculation of the stock of tangible capital (see, for instance, Gehringer, 2013). Accordingly, the stock of patents in each year t in country i is given by:

$$STOCK_{it} = (1 - \delta)STOCK_{it-1} + FLOW_{it} \qquad (7A.1)$$

where $FLOW_{it}$ measures the current flow of patents and δ is a depreciation rate assumed constant and equal to 0.06. Moreover, the initial level of the patent stock, $STOCK_{i0}$ is computed as:

$$STOCK_{i0} = FLOW_{i0}/g + \delta \qquad (7A.2)$$

where g is a geometric average of the growth rate of investment over the whole period for which data on each year patent applications are available. For the majority of countries in our sample this was between 1963 and 2011, with the exception of Australia (1995–2011), Belgium (1965–2011), Canada (1960–2011), Korea (1960–2011), the Netherlands (1960–2011) and Spain (1965–2011).

8. Productivity growth persistence: firm strategies, size and system properties

Cristiano Antonelli, Francesco Crespi and Giuseppe Scellato

1. INTRODUCTION

Over the past few decades a broad range of research activities has been dedicated to the study of productivity growth and its sources. Traditionally, empirical analyses were based on macro- or industry-level aggregate data. More recently, a large number of studies based on micro data has been produced also due to the increasing availability of firm-level data (for extensive reviews, see Bartelsman and Doms, 2000; Ahn, 2000; Foster et al., 2001; Syverson, 2011; Mohnen and Hall, 2013). The discovery of ubiquitous, extensive and persistent productivity differences has shaped research agendas in a number of fields. Macroeconomists break down aggregate productivity growth into various micro-components with the aim of providing a better understanding of the sources of this growth. Models of economic fluctuations driven by productivity shocks are increasingly being enriched to account for micro-level patterns, and are estimated and tested using plant- or firm-level productivity data (Bartelsman et al., 2009). In this context it has been possible to analyse the differentiated role played by firms of different size groups in explaining aggregate patterns of productivity growth. This has led to recognizing the contribution that SMEs may have in fostering productivity growth due to the process of creative destruction that they are able to engender (Hölzl, 2009; Henrekson and Johansson, 2010; Colombelli et al., 2014) and to their ability to creatively adapt existing technological knowledge to the conditions of local product and factor markets (Antonelli and Scellato, 2015).

Moreover, two important lessons have been learned from this extensive field of research. First, the level of productivity dispersion is extremely large, that is, some firms are markedly more efficient than others. Second, firms that are highly productive today are more likely to be highly productive tomorrow. In other words, the literature has clearly pointed out the existence of a high degree of persistence in productivity differences across producers (Bartelsman and Doms, 2000; Syverson, 2011; Raymond et al., 2013).

The identification of such high and yet persistent productivity dispersion across producers has led to the emergence of a huge amount of empirical literature that attempts to explain the sources of these productivity patterns. This evidence casts major doubts on and raises substantial criticism of the new growth theory according to which the rates of productivity growth and of the introduction of technological innovations should be homogeneous across firms that belong to the same system (Aghion and Howitt, 1998). The relevance of this empirical evidence and its theoretical implications has led to the identification of a number of factors that could determine systematic differences in the productivity performances of producers, including the role of innovative activities and the diffusion of ICTs (Griliches, 1979; Brynjolfsson and Hitt, 2000; Crépon et al., 1998; Faggio et al., 2009 Raymond et al., 2013). Among firm-specific characteristics that are capable of affecting the productivity growth of producers, particular attention has been devoted to assessing the impact of human capital and the quality of management practices on different measures of productivity and firm performance (for recent contributions, see McEvily and Chakravarthy, 2002; Ilmakunnas et al., 2004; Galindo-Rueda and Haskel, 2005; Bou and Satorra, 2007; Bloom and Van Reenen, 2007, 2010).

In this context of analysis, the presence of persistent patterns of above-average productivity growth at the firm level can be interpreted as the result of the ability of firms to exploit dynamic capabilities to sustain competitive advantages, as highlighted by management studies (Teece and Pisano, 1994; Verona and Ravasi, 2003; Teece, 2007; Vergne and Durand, 2011). In particular, building on recent developments in the economics of innovation that have paid attention to the analysis of innovation persistence (Malerba et al., 1997; Cefis, 2003; Peters, 2009; Roper and Hewitt-Dundas, 2008; Antonelli et al., 2012, 2013), it can be claimed that the repeated interactions between the accumulation of knowledge and the creation of routines to valorize and exploit it within the same organization may lead to the creation of dynamic capabilities that favour the systematic realization of above-average productivity (Nelson and Winter, 1982; Rothaermel and Hess, 2007; Verona and Ravasi, 2003).

This framework emphasizes that the past has a significant impact on current and future performance. However, the dynamic capabilities approach recognizes that a business enterprise is shaped but not necessarily trapped by its past. Management strategies can make big differences through investment choices and other decisions. Hence, the role of knowledge cumulativeness and the relevance of strategic decisions to leverage internal and external knowledge are considered to be crucial in shaping path-dependent dynamics of productivity growth (Antonelli et al., 2013; Crespi and Scellato, 2014).

Moreover, in the present study, we investigate the potential role of size as another important firm-level characteristic capable of shaping persistent patterns of productivity growth. On the one hand, large corporations may have an advantage in sustaining higher performance in terms of productivity growth for longer time spans due to their superior ability to invest in R&D activities and benefit from high levels of cumulated knowledge (Chandler, 1977, 1990). On the other hand, persistency patterns can be independent of size, as shown by the literature on 'gazelles' – that is, high-growth firms where persistent abnormal sales or employment growth rates have been identified in subsets of companies belonging to all size classes (Henrekson and Johansson, 2010; Lopez-Garcia and Puente, 2012; Colombelli et al., 2014).

Finally, the proposed analysis aims to take into account the role played by system properties that shape the context in which the persistence of total factor productivity (TFP) growth occurs. In particular, the effects on persistency patterns played by the amount of knowledge externalities, the dynamics of market forces and the different types of sectoral systems are explored.

The empirical analysis is based on a large sample of Italian firms and follows a two-step empirical strategy consisting of a preliminary identification of persistence in TFP growth through transition probability matrices (TPMs) and an econometric analysis that aims to qualify the persistence of productivity growth as an emergent system property that depends on the combination of firms' characteristics and specific properties of the system in which the strategy of firms take place. This chapter is structured as follows: the literature on persistence in productivity is reviewed in Section 2; the hypotheses and research design of this study are outlined in Section 3; empirical evidence is presented in Section 4; and the main results are summarized in the conclusions.

2.   THE PERSISTENCE OF PRODUCTIVITY PERFORMANCE

Under the assumption of random productivity differences across producers, relative productivity would be uncorrelated from one period to another.

There would be no persistence in productivity distribution, and the TFP of a producer in one period would have no predictive power on the TFP in another period. However, empirical investigations have shown that there are large and persistent differences in productivity across plants and firms in the same industry (Bartelsman and Doms, 2000). When analysing persistence in productivity, many studies have followed an approach based on TPMs relative to plant/firm productivity distribution (see, for example, Baily et al., 1992; Bartelsman and Dhrymes, 1998). The calculated transition matrices exhibit large diagonal and near-diagonal elements, indicating that producers that are high in the distribution in one period tend to continue to have a high rank in the distribution in subsequent periods.

Baily et al. ranked the plants in their sample for 1972–1988 according to their relative productivity for each year and divided them into quintiles. They then calculated a transition matrix that highlighted 'an enormous amount of persistence in the productivity distribution' (Baily et al., 1992: 219). Of all the plants that were in the first quintile in 1972, a weighted 60.75 per cent were again in the first quintile in 1977 and, of all the plants that were in the first quintile in 1977, a weighted 52.89 per cent had come from the first quintile in 1972. The persistence in the 10-year transitions was even stronger than that found for five years. More than 58 per cent of the plants in the top quintile in 1972 were still in the top two quintiles in 1982. Bartelsman and Dhrymes found a similar high degree of persistence in productivity ranking through an examination of the behaviour of TFPs in selected industries over the 1972–86 period in the USA. They showed, in particular, that about 60 per cent of the plant-year observations did not move away by more than one decile from their previous rank. Moreover, they found that larger plants exhibited more stability and that the probability of staying close (one decile) to the previous position increased with age and size. They concluded that this evidence could have been the result of some form of 'learning by doing' that may characterize the evolution of the productivity performance of plants.

More recently, Giannangeli and Gomez-Salvador (2008) have used annual account data over the 1993–2003 period for a balanced panel of manufacturing firms in a selected panel of five European countries. They found a high degree of persistence of the relative efficiency of firms. Around 25 per cent of firms in all countries considered in the analysis remained in the middle of the distribution, while more than half of the sample persistently remained at the top and bottom. The authors concluded that the high persistence of relative productivity levels suggests that firm-efficiency levels are structurally different from firm to firm.

As far as Italy is concerned, Bottazzi et al. (2008) carried out an analysis based on a large panel of Italian firms active in both manufacturing and services during the 1998–2003 period that confirmed the presence of a strong and positive correlation in productivity over time. They explored the links between the persistence in productivity and profitability, and found that more-efficient firms also tend to be more profitable.

Although these empirical investigations have shown that there are large and persistent differences in productivity levels across plants and firms, productivity growth rates have usually been found to exhibit an important transitory component. Baily et al. show clear evidence of regression to the mean effects in productivity growth regressions. Similarly, Bartelsman and Dhrymes detected a strong negative correlation between a plant's growth rate over a five-year period and its productivity growth over the previous five years. Giannangeli and Gomez-Salvador, on the other hand, showed that when lagged productivity growth is included in the econometric model, it is positive and significant, thus indicating some persistence in labour productivity growth at the firm level.

Geroski et al. (2003, 2009) specifically investigated persistence in productivity growth. In their first paper, using a sample of 147 UK firms observed continuously for more than 30 years, they showed that growth rates are highly variable over time and that the differences in growth rates between firms do not persist for very long. This outcome was considered to be due to the random nature of the innovative activities of firms which translates into random shocks on productivity. Again, in the second paper, they found that, in general, individual firms do not outperform their peers for very long when stable firm characteristics, via firm-fixed effects, are accounted for. However, the analysis showed that the few instances of sustained productivity growth performance that had been observed appeared to have been triggered mainly by prior innovative activity and the disciplining effect of corporate debt.

The significance of the role of innovation in determining the persistence of productivity performance can be better understood by recognizing that innovation itself is characterized by a certain degree of persistence. The theme of innovation persistence in recent years has attracted the interest of scholars in different research perspectives, ranging from the economics of knowledge to the economics of organization and the economics of innovation (Malerba et al., 1997; Cefis and Orsenigo, 2001; Peters, 2009; Antonelli et al., 2012, 2013; Clausen et al., 2012).[1] Most of the empirical analysis provides evidence in favour of the presence of persistency patterns in innovative efforts at the firm level related to technological learning processes that eventually generate new knowledge for the innovating company. As suggested by Geroski et al. (2003, 2009), these effects probably translate into the dynamics of TFP growth at the firm level.

3. THE SOURCES OF PERSISTENCE IN PRODUCTIVITY GROWTH

The reviewed evidence indicates that productivity growth persistence may be substantial, suggesting that past productivity performance influences subsequent patterns. However, it seems clear that previous behaviour is not sufficient to warrant the ability to keep outperforming levels of total factor productivity. TFP growth persistence occurs when a number of complementary and contingent factors sustain and strengthen the hysteresis generated by previous dynamics. The identification of the dynamic features of the sources of persistence is at the core of our analysis. In this respect, a firm's TFP reflects the levels of a broad range of technological, organizational and managerial capabilities, along with the ability to exploit them through appropriation of the results of introducing technological innovations. TFP performance is in fact related to the systematic ability to generate new knowledge, apply it to the broad array of activities that firms carry out and to exploit it. The exploitation of knowledge includes both the introduction of innovations and the adoption of technological and organizational innovations introduced by suppliers and competitors.

Knowledge cumulability, related to knowledge indivisibility and knowledge non-exhaustibility, plays a central role in this context. The achievement of higher performance in terms of productivity dynamics can be easier for firms that can command a larger internal knowledge base and have access to and the ability to use larger knowledge bases than other agents operating in the same system (Antonelli, 2011; Colombelli and von Tunzelmann, 2011). For this reason, the effects of knowledge cumulability are typically path dependent (David, 1985, 2007).

Knowledge accumulated in the past exerts a strong influence on the future generation of new knowledge. Such effects, however, can change over time because the rates of accumulation and the conditions of access are not fixed (Dobusch and Schüler, 2013). Past knowledge, in fact, is not the single, deterministic factor: management strategies appear to be crucial in shaping the amount of knowledge that each firm is able to generate at each point in time and in sustaining persistent higher performance in terms of TFP growth through R&D investment choices and other decisions related to the acquisition of specific pieces of external knowledge. In this respect, the economics of organization has shown that repeated interactions between the accumulation of knowledge and the creation of routines to valorize and exploit it within the same organization eventually lead to the creation of dynamic capabilities that favour the systematic reliance on innovation as a competitive tool (Nelson and Winter, 1982; Rothaermel and Hess, 2007; Verona and Ravasi, 2003). In particular, only firms able to leverage their dynamic capabilities can be persistently among the top TFP performers over a long period of time (Teece, 2007).

This framework emphasizes that the past has an important impact on current and future performances. However, as acknowledged by the dynamic capabilities approach, firm strategies are indeed influenced but not necessarily trapped by their past. Management can make big differences through investment choices and other decisions. Hence, managers can act creatively and strategically to shape firms' growth paths (Parker et al., 2010). Heterogeneity in the form of strategic differences across firms constitutes a key driving force behind a firm's probability of sustaining TFP growth over time (Clausen et al., 2012). In this perspective, firms focusing their strategy on acquisition, assimilation and exploitation of externally available knowledge are able to continuously renew their knowledge stock and strengthen their dynamic capabilities. Hence, managers can deal with and even reap benefits from path dependence if they are able to select appropriate self-reinforcing mechanisms along the capability paths that emerge from the firm–environment interaction (Vergne and Durand, 2011). Therefore, management strategies appear to be crucial to sustaining superior productivity performance over time through investment choices and other decisions related to the leveraging of dynamic capabilities and the exploitation of strategic assets. Managerial contingencies in fact affect the non-ergodic dynamics of innovation persistence (Clausen et al., 2012).

In this analysis it is argued that productivity persistence is an emergent system property that takes place when there is appropriate matching between the system properties and the characteristics and conduct of individual agents. This amounts to specifying the hypothesis that productivity persistence is determined by a mix of strategic decisions, firm-level characteristics with special reference to size and system properties. Since the dynamics of emergent system properties are influenced by a mix of interacting factors where each one exhibits high levels of dynamic variance and non-ergodicity, path dependence is claimed to be an intrinsic feature of TFP growth persistence. The interaction between different processes is in fact most likely to generate path rather than past dependence. Hence we put forward the complementary hypothesis that the persistence of productivity exhibits the typical traits of a non-ergodic process influenced by the past and yet sensitive to events that occur along the growth path.

In order to capture the effects related to firms' managerial strategies on persistency patterns over time we have focused our attention on three main dimensions: the decisions related to business process outsourcing; the strategies on the accumulation of intangible assets; and the propensity to assume a long-term perspective in investment choices. Analysis of these aspects should allow us to qualify the observed firms in terms of their strategic commitment to rely on the valorization of intangible assets and dynamic capabilities to persistently sustain superior productivity performance.

With respect to the first dimension, the literature suggests that in the last decades there has been an increasing tendency by firms to outsource a significant part of their non-core activities in order to achieve advantages in terms of productivity increases (Gilley and Rasheed, 2000; Grossman and Helpman, 2005; Merino and Rodríguez, 2007; Amiti and Wei, 2009). When outsourcing processes are successfully implemented, firms are able to focus on their core competences and hence improve efficiency and productivity, explore new potential sources of revenues and implement new investment projects (Heshmati, 2003; Broedner et al., 2009). Moreover, through business outsourcing, external long-run sources of TFP growth may be activated, as those services are typically carried out by highly specialized experts and heavily rely on ICT, a major driver of productivity gains (Abramovsky and Griffith, 2006; Crespi, 2007). Finally, significant complementarity effects between internal and external R&D may emerge when business processes are successfully outsourced (Lokshin et al., 2008).

In parallel, the strong heterogeneity in firms' investments in intangible assets and the identification of strong cumulative processes of intangible asset accumulation suggests that the different propensity to invest in intangible assets can be explained by specific characteristics, internal capabilities and managerial strategies at the firm level (Arrighetti et al., 2014). In this context, we expect persistence in higher productivity performance to be sustained when managers adopt a strategy based on the systematic reliance on and valorization of intangible assets as ways to leverage dynamic capabilities (Bontempi and Mairesse, 2008; Corrado et al., 2009; Marrocu et al., 2012).

Finally, intangible investment, in particular that related to R&D activities, tends to be approached as a long-term investment whose influence, in terms of business performance, is shaped by experience and learning processes provided by previous accumulation of technological, organizational and other capabilities (Winter, 1987). Hence, we expect the financial structure of firms to reflect the strategic commitment of managers to long-term investment. Hence, a higher propensity to rely on long-term debt can be interpreted as an indication of a strategic perspective to continuously fuel competitive advantages.

In addition to strategic factors, other internal characteristics of firms are expected to play a role. Firm size is a specific internal characteristic that is worth analysing in the context of productivity persistence. In particular, we claim that it is important to distinguish between two effects of company size on the dynamics of productivity. The first is a direct influence due to the relation of size to various efficiency-enhancing activities such as the use of ICT, labour skills and training activities, the intensity of R&D investments and the introduction of innovations (Cohen and Klepper, 1996; Bartelsman and Doms, 2000; Syverson, 2011). The second may be related to the idea that cumulability effects are mainly relevant in large corporations: the hysteretic influence of past productivity growth on current and future performance increases with size, along with the accumulation of competence as a strategic asset.

However, while we expect the first effect of size to be relevant, we suppose that persistency patterns can be identified independently of size. Small firms can counterbalance the effects of knowledge cumulability that favour corporations with positive effects of entrepreneurship (Colombelli et al., 2014). Moreover, recent empirical evidence showed that persistence in innovation activities can be found also in the case of SMEs, where persistency appears to be shaped by success breeds success, sunk costs and demand-pull effects (Máñez et al., 2014; Le Bas and Scellato, 2014), with a stronger association between sales growth and subsequent R&D growth in small firms than larger firms (Deschryvere, 2014). Such persistency patterns in SMEs can be particularly relevant in high-tech industries (Máñez et al., 2014), where the role of technological start-ups is potentially significant (Santarelli and Vivarelli, 2007) and, in general, has been found to represent an important source of employment and productivity growth (Triguero et al., 2014).

Finally in this analysis we highlight the role of system properties such as the amount of knowledge externalities and the dynamics of market forces. As the economics of knowledge suggests, different forms of external knowledge (that is, scientific, commercial, technological and organizational), as well as different kinds of activities close to R&D activities and learning (such as searching, networking, absorption and scientific outsourcing), are required to generate and exploit new knowledge (Adams, 2006; Johansson and Lööf, 2008). Following this approach, the system properties add to internal ones and shape the context in which the persistence of TFP growth occurs. This approach is confirmed by a wave of recent empirical studies that stress the role of system properties in shaping the strategies of firms that are able to be persistent innovators, namely Ito and Lechevalier (2010), who stress the positive role exerted by qualified interactions in international markets with qualified users for Japanese firms that are able to export systematically.

This positive feedback supports the persistent introduction of innovations as a key component in firms' exporting strategies. Triguero and Corcoles (2013) and Triguero et al. (2014) emphasize the role of technological opportunities, appropriability conditions and market demand, arguing that Spanish firms that take advantage of these factors are better able to implement the persistent introduction of innovations. Suarez (2014) confirms the important role of the macroeconomic context, showing that persistence is stronger in economic systems that grow faster and have lower levels of instability. Along similar lines, Bergek and colleagues (2013) provide rich empirical evidence in the automotive and gas turbine industries confirming that incumbents are able to implement the persistent introduction of innovations based on the 'creative accumulation' stirred by the entry of new competitors and technological discontinuities brought about by new technological opportunities. The evidence on Korean firms analysed by Kim and Chang-Yang (2011) confirms that early entrants in industries characterized by high levels of knowledge appropriability and low levels of technological opportunities are able to implement their innovative leadership and become persistent innovators retaining and replicating first-mover advantages.

In sum, a wave of recent studies confirms that the persistence of technological leadership is the result of dynamic capabilities implemented by typical Schumpeterian strategies of firms that are able to expand their knowledge base and use it as a tool to retain the competitive advantage based on the introduction and adoption of innovations in the past, provided that the system properties are favourable. As Brian Loasby (2010) puts it, capabilities deliver effective performances only in a specific context that includes aspects that are both internal and external to each firm.

Hence, following the distinction elaborated by Harper and Lewis (2012) between resultant properties that qualify both individual agents and aggregate and emergent system properties where individual properties are qualified by the characteristics of the system in which they are embedded, we can put forward the hypothesis that the persistence of innovation is an emergent system property, rather than a resultant property. This hypothesis stems directly from the legacy of Schumpeter (1947), which explains the introduction of innovations as the result of the creative reaction of firms made possible by entrepreneurial strategies contingent upon the system properties. Specifically, this chapter stresses the role of system properties together with the characteristics and strategies of firms, in particular the level of dynamic capabilities, as critical factors that make the persistent achievement of superior TFP growth possible. In this context, the persistence of productivity growth acquires all the characteristics of an emergent property of the system in which firms are embedded that shares the typical traits of a path-dependent process, influenced by the past and yet sensitive to events that occur along the growth path (Antonelli, 2011).

4.   EMPIRICAL ANALYSIS

In order to test the relevance of these arguments, a two-step empirical strategy was set up. In the first step the analysis focused on the identification of persistence in TFP growth through a sequence of transition probability matrices (TPMs) considering different sub-samples. Such an approach accounts for changes that take place throughout the process which are expected to have significant effects on the path-dependent dynamics of TFP persistence (Antonelli, 1997). In the second step, the analysis concentrated on the drivers of persistence in order to qualify the role of the contingent events that affect the dynamics at work.

4.1   Dataset

The dataset is based on financial accounting data from a large sample of Italian manufacturing companies observed over the years 1996–2005. The original data were extracted from the Aida database provided by Bureau van Dijk, which reports complete financial accounting data for public and private Italian firms with a turnover greater than 0.5 million euros. The companies included in the analysis were founded before 1995, were registered in a manufacturing sector according to the Italian ATECO classification, and were still active by the end of 2005. All companies with at least 15 employees at the end of the 1995 fiscal year were included. After collecting balance sheet data, all the companies with missing values were dropped. In order to drop outliers, due to possible errors in the data source, we computed a number of financial ratios and yearly growth rates of employees, sales and fixed capital stock. After manual checking we eventually dropped 45 companies, and thus ended up with a balanced panel of 7020 companies. All financial data were deflated according to a sectoral three-digit deflator using year 2000 basic prices. For the analysis of the sectoral distribution of the companies see Antonelli et al. (2015).

The firm-level TFP was calculated using Cobb–Douglas production functions with constant return to scale for each industry included in the sample:

$$TFP_{i,t} = \frac{Q_{i,t}}{L_{i,t}^{\beta} k_{i,t}^{1-\beta}} \tag{8.1}$$

where $Q_{i,t}$ is deflated value added, $L_{i,t}$ average number of employees and $K_{i,t}$ fixed capital stock.

In order to compute the capital stock through time a perpetual inventory technique was applied according to which the first year accounting data (that is, 1996 in the present case) were used as the actual replacement values. The subsequent yearly values of fixed capital were computed using a depreciation parameter δ, assumed equal to 6.5 per cent, and adding deflated yearly investments.[2] The investment parameter $(I_{i,t})$ was computed as the yearly variation in the net fixed capital in the companies' balance sheets plus yearly amortizations. Hence, the time series of fixed capital is defined as:

$$K_{i,t} = (1 - \delta)K_{i,t-1} + I_{i,t}/p_t \tag{8.2}$$

In order to identify the parameter β at industry level to compute equation (8.2), the following equation was estimated for each industry, where α_i is a firm-specific effect and α_t a time-specific effect:

$$Log\left(\frac{Q_{i,t}}{k_{i,t}}\right) = \beta \times Log\frac{L_{i,t}}{k_{i,t}} + \alpha_i + \alpha_t + \varepsilon_{i,t} \tag{8.3}$$

In order to analyse the dynamics of firm-level TFP growth rates we calculated the variable ΔTFP, defined as the logarithmic growth rate of the TFP level between year $t - 3$ and year t:

$$\Delta TFP_{i,t} = \log(TFP_{i,t}) - \log(TFP_{i,t-3}) \tag{8.4}$$

We then proceeded with a classification of the values taken from the variable $\Delta TFP_{i,t}$ on the basis of the distribution of the TFP growth rates of all the companies in the same sector of company i between year $t - 3$ and year t. This procedure allows us to evaluate the persistence of firm-level TFP growth rates, taking into account industry-specific trends. In particular, we analyse the probability of a company's TFP growth rate being persistently located within a specific quantile of the distribution of TFP growth rates of all companies in the same industry.[3] Sensitivity analyses were conducted to assess whether, and to what extent, the thresholds adopted for the discretization of the TFP growth rate distribution (for example, using tertiles or quartiles) affect the estimated intensity of persistence.

Two complementary approaches were followed in the empirical analysis. Initially, we investigated the presence of firm-level persistence by means of transition probability matrices (TPM). Then, we explored firm-level persistence by means of discrete choice panel data models, based on the estimator proposed by Wooldridge (2005). While the initial TPM approach is expected to provide only summary evidence on the persistence of the TFP growth rates of firms over time, the panel data analysis is aimed at identifying true state persistence after controlling for relevant contingent factors. Independent variables used in the econometric analysis include size, return on equity, leverage, an indicator of vertical integration, an indicator of debt maturity composition and intangible intensity, computed as the yearly incidence of intangible to tangible

Table 8.1   Description and summary statistics of the variables used in the econometric analysis

Variable	Description	Mean	Median	Std dev.	1%	99%
SIZE i,t	Log of total assets of company *i* in year *t* (based on the perpetual inventory method)	14.30	14.33	1.38	10.97	17.70
INTANG i,t	Ratio of book values of intangible assets to tangible assets for company *i* in year *t*	0.15	0.08	0.19	0	0.85
LEV i,t	Book value of debt/ (book value of debt + book value of equity)	0.68	0.72	0.20	0.17	0.98
ROE i,t	Net income/book value of equity	0.32	0.04	0.6	−1.59	0.73
VERT_INT i,t	Value added/turnover	0.28	0.28	3.30	0.05	0.68
DEBT_MAT i,t	Long-term debt/total debt	0.13	0.08	0.15	0	0.61
EMPLOYEES i,t	Number of employees	111	56	330	16	921

assets. Table 8.1 reports summary statistics of the variables used in the econometric analysis.

4.2   Transition Probability Matrices on TFP Growth Rates

The following three tables report the results obtained for the persistence of TFP growth rates over time, using different discretization criteria. In Table 8.2 we calculated the TPM by splitting the distribution of firm-level TFP growth rates into tertiles. We also report the standard errors of the related transition probabilities in the table.[4]

The data show that, during the observed years, the firms that were in the top tertile of TFP growth rates in their sector in year *t* – 1 were again in the top tertile in year *t* with a probability of 54.04 per cent. Overall, the data in Table 8.2 highlight the presence of strong persistence: the main diagonal terms are larger than 33 per cent. The incidence of inter-temporal transition between the lowest and the highest tertile is quite low in both directions and is below 20 per cent. The analysis was replicated by splitting the distributions into quartiles (see Appendix Table 8.A1). Again, the data confirmed the presence of non-negligible persistency patterns. As could be expected, inter-quantile mobility was higher for the intermediate intervals. This evidence seems to highlight the presence within the sample of subpopulations of firms that are capable of repeatedly outperforming their peers in terms of TFP growth.

Interestingly, persistence in TFP growth rates is a phenomenon which appears not to be confined to large companies, as shown in Table 8.3,

Table 8.2 *TPM on the tertiles of sectoral distribution TFP growth rates for all years and all companies*

	High Growth t	Mid Growth t	Low Growth t
High Growth $t-1$	0.5404	0.2776	0.172
	(0.0041)	(0.0035)	(0.0031)
Mid Growth $t-1$	0.2911	0.4232	0.2857
	(0.0038)	(0.0041)	(0.0038)
Low Growth $t-1$	0.1807	0.2826	0.5367
	(0.0032)	(0.0038)	(0.0042)

Note: Standard errors in parenthesis.

Table 8.3 *TPM on the tertiles of sectoral distribution TFP growth rates for all years (firms split according to size)*

	High Growth t	Mid Growth t	Low Growth t
Firms with more than 250 employees at $t-1$			
High Growth $t-1$	0.5746	0.2625	0.1569
	(0.017)	(0.015	(0.012)
Mid Growth $t-1$	0.2730	0.4378	0.2892
	(0.014)	(0.016	(0.014)
Low Growth $t-1$	0.196	0.2607	0.5457
	(0.011)	(0.013	(0.014)
Firms with less than 250 employees at $t-1$			
High Growth $t-1$	0.5341	0.2836	0.1823
	(0.004)	(0.004)	(0.003)
Mid Growth $t-1$	0.2867	0.4241	0.2913
	(0.004)	(0.004)	(0.004)
Low Growth $t-1$	0.1700	0.2901	0.5399
	(0.003)	(0.004)	(0.004)

Note: Standard errors in parenthesis.

though the transition probabilities in the main diagonal of the matrices are greater in the group of large companies. However, this last result may be spurious as it may simply reflect the relevance of what we called the direct effect of size on productivity dynamics. A better assessment of this issue will be provided through the econometric analysis.

In Tables 8.4 and 8.5 we split the transition probability matrices, considering different sub-periods, sectors and regions. This splitting approach has the aim of capturing the presence of divergences in persistency patterns of TFP growth rates due to the influence of system properties. In particular, we claim that the knowledge intensity of the local context may be relevant in shaping differentiated patterns of persistence in productivity growth. For this purpose, we split Italian regions into High R&D and Low R&D regions on the basis of average aggregate R&D expenditures during the observed years. High R&D regions fall into the top 33 per cent of distribution of regions in terms of gross R&D expenditures/GDP. Moreover, the macroeconomic context is expected to play an important role in influencing the productivity performance of firms and their reactions to changing economic conditions in terms of contingent behaviour and strategic decisions.

Considering that the time span adopted for the analysis can be conveniently divided into two sub-periods that identify an upward economic cycle (until 2001) and a downward cycle (after 2001) in Italy, we split the TPMs in order to eventually detect any differences in persistency dynamics across the two sub-periods. Finally, we divided the sample into groups according to the technological intensity of different economic sectors in order to account for the effects that sectoral system properties may have on the persistence of TFP growth.[5] Table 8.4 reports figures on the share of companies belonging to the top 15 per cent of the sectoral distribution of TFP growth rates for two subsequent periods, for subsamples selected according to periods, the R&D intensity of regions and the technology intensity of sectors.[6]

This approach allows us to see whether the observed aggregate persistency patterns are the averaged outcome of processes with peculiar trends over different regions, sectors and time. The differences between the results can in fact be interpreted as a first indication that system properties are capable of shaping persistence by affecting its dynamics.

The results reported in Table 8.4 show that there are significant differences between the different sub-groups, ranging from 0.32 to 0.45. The highest transition probability for top-performing companies in two subsequent periods is associated with the group of firms in High R&D regions, and Hightech sectors observed in 2001–2005.

The role of sectoral systems appears to be of particular relevance since,

Table 8.4   Share of companies in the top 15% of the sectoral distribution of TFP growth rates in two subsequent times for selected regions, sectors and periods

Region	Sector	Period	Probability of high-high growth	Standard error
LOW R&D	HITECH	1998–2000	0.4040	0.0236
LOW R&D	HITECH	2001–2005	0.3829	0.0183
LOW R&D	LOWTECH	1998–2000	0.3671	0.0199
LOW R&D	LOWTECH	2001–2005	0.3758	0.0135
HIGH R&D	HITECH	1998–2000	0.3718	0.0232
HIGH R&D	HITECH	2001–2005	0.4514	0.0169
HIGH R&D	LOWTECH	1998–2000	0.3224	0.0173
HIGH R&D	LOWTECH	2001–2005	0.3906	0.0133

everything else being equal, the probability of continuously performing in the top 15 per cent of the distribution is always higher for firms operating in High-tech sectors. In parallel, the macroeconomic cycle appears to have a differentiated impact according to the type of region where business takes place. In particular, for High R&D regions there is a significant tendency towards an increase in TFP persistence after 2001, while the dynamics is more stable for Low R&D regions.

When interpreting these results, it is fundamental to consider that companies that start with lower TFP levels are more likely to exhibit higher TFP growth rates. This aspect could be relevant in explaining the high level of persistence for outperforming companies in Low R&D regions before 2001, which may be related to firms starting with lower TFP levels and taking advantage of the macroeconomic expansion that occurred until 2001. Inversion of the economic cycle has powerful effects: firms belonging to the High R&D regions seem to be more capable of sustaining persistently higher levels of productivity growth. One possible interpretation of this result is that in this economic phase the dominating effect could be related to the best companies that react to the changed economic context and which strategically invest in innovative activities to sustain persistently higher TFP gains.

4.3   Econometric Analysis

4.3.1   Modelling structure
The previous descriptive evidence clearly calls for a more detailed analysis of the actual underlying dynamics and its driving factors. In order to

analyse the persistence of TFP growth rates throughout the studied periods, we constructed a time-varying dummy variable that equals one in period t if a company shows a TFP growth rate that falls within the top 15 per cent of the distribution of ΔTFP for all the companies in the same sector. We applied a dynamic discrete choice model in which such a variable is regressed against its past realization and a set of appropriate controls.[7]

The observed persistence may be due to true state dependence or permanent unobserved heterogeneity across the analysed companies. From a theoretical perspective, if the source of persistence is due to permanent unobserved heterogeneity, individuals show higher propensity to make a decision; but there is no effect of previous choices on current utility, and past experience has no behavioural effect (Heckman, 1981). Hence, in order to estimate true state persistence, it is important to capture the variance of the state indicator which is explained by both the structural characteristics of the firms and by contingent, time-varying observable factors and then to analyse whether its past values still have a significant effect (Peters, 2009; Antonelli et al., 2012; Lopez-Garcia and Puente, 2012).

The baseline specification for a dynamic discrete response model is the following:

$$y^*_{it} = \gamma y_{it-1} + \beta x_{it} + u_i + \varepsilon_{it} \tag{8.5}$$

where y_{it} (with possible values 0,1) is the state indicator (that is, indicating whether a firm is in the top 15 per cent of the TFP growth rate distribution in its sector in year t).

The estimation of the above model is based on a strong assumption on the initial observations y_{i0} and their relationship with u_i, the unobserved individual effects. If the origin of the analysed process does not coincide with that of the available observations, y_{i0} cannot be treated as exogenous, and its correlation to the error term would give rise to biased estimates of the autoregressive parameter that represents the measure of persistence. In order to deal with this issue in this chapter we apply the method developed by Wooldridge (2005), which proposes specifying the distribution of u_i conditional on y_{i0} and x_i. In particular, we follow the approach used by Peters (2009) by using the first realization of the dependent variable (y_{i0}) and the time-averaged covariates as predictors of the individual effect.

As previously mentioned, in order to identify true state persistence, it is necessary to account for the time-varying firm-level characteristics which are expected to be correlated to the observed outcome of the dichotomous dependent variable, and to control for the properties of the system in which firms operate. With respect to aspects related to firms' strategies, we tested the relevance of three variables that are linked to management decisions to sustain persistent higher productivity growth rates – that is, the indicator of vertical integration along the value chain (VERT_INT), the indicator of debt maturity composition (DEBT_MAT) and the indicator of intangible intensity (INTANG). As highlighted in Section 3, we claim that lower values of vertical integration can be attributed to the strategic decision of focusing on those segments of the value chain that are characterized by higher value added. Hence, a negative relationship with TFP growth rates is expected.[8]

Moreover, while in principle the capital structure should have a neutral or non-significant effect, it is here claimed that a higher incidence of long-term debt can be associated with the willingness of managers to adopt a more long-term investment strategy. Since structural innovation investments require a stable commitment, we expect sustained higher performances in TFP growth rates to be observed for those firms that have longer debt maturity. This, in turn, is a signal that such firms have made significant investments in long-term infrastructures. Finally, intangible assets intensity is expected to capture the effort of a firm to build innovative competences by means of both in-house R&D and external expenditures.

Additional firm-level variables that are likely to affect TFP growth used in the econometric analysis include firm size (SIZE), leverage (LEV) and an indicator of firm profitability, return on equity (ROE).[9] With respect to the role of firm size, besides taking into account the direct effect of size on productivity performances by including this variable in the tested models, we will also test if differences in the impact of previous TFP growth rates on current performances emerge for groups of firms of different size. All time-varying firm-specific factors have been used in the model specifications with a three-year lag.

Finally, in the analysis we take into account the effects of the properties of the system by testing our model after splitting the sample of companies according to the categories used in the descriptive analysis – namely, two sub-periods (before and after 2001), two macro-regions (High R&D and Low R&D) and macro-sectors (High-Tech and Low-Tech industries). The time splitting should help us grasp the effects on persistence of the macroeconomic cycle. The regional splitting should capture the effects of the local intensity of R&D expenditures, while the sectoral splitting should account for the role of sectoral systems in productivity persistence. Consistently with the review of the literature and the set of hypotheses that we have developed, we expect to find systematic and significant differences between the sub-samples that indicate that the properties of the system affect the strategy of firms, with consequent effects on persistence.

4.3.2   Results

In the following tables we show the results of the model tested on different samples for the evaluation of persistence along time of TFP growth rates. Results reported in Table 8.5 (column I) confirm the summary evidence

Table 8.5   Dynamic probit model on the persistence of TFP growth rates

	All years	1998–2000	1998–2000	2001–2005	2001–2005
	All sectors	High-Tech	Low-Tech	High-Tech	Low-Tech
	I	II	III	IV	V
HighGrowth $_{t-1}$	0.269***	0.232***	0.112***	0.309***	0.271***
	(0.007)	(0.047)	(0.030)	(0.013)	(0.010)
SIZE $_{t-3}$	0.072***	0.069***	0.081***	0.057***	0.083***
	(0.004)	(0.011)	(0.008)	(0.009)	(0.008)
ROE $_{t-3}$	0.000	−0.018***	−0.009***	0.001	0.000
	(0.000)	(0.005)	(0.002)	(0.001)	(0.000)
INTANG $_{t-3}$	0.014	−0.015	0.060*	−0.023	0.026
	(0.017)	(0.045)	(0.033)	(0.032)	(0.027)
VERT_INT $_{t-3}$	−0.270***	−0.187***	−1.006***	−0.316***	−0.298***
	(0.013)	(0.032)	(0.064)	(0.026)	(0.024)
LEVERAGE $_{t-3}$	0.001	−0.007	−0.032***	0.010	0.014
	(0.001)	(0.006)	(0.012)	(0.006)	(0.010)
DEBT_MAT $_{t-3}$	0.096***	0.114**	0.065**	0.060*	0.105***
	(0.017)	(0.046)	(0.032)	(0.033)	(0.029)
AVGSIZE	−0.068***	−0.058***	−0.081***	−0.053***	−0.079***
	(0.005)	(0.011)	(0.008)	(0.009)	(0.008)
AVGROE	0.001*	0.009**	−0.000	0.002	0.001
	(0.000)	(0.004)	(0.001)	(0.002)	(0.001)
AVGINTANG	0.071***	0.041	−0.038	0.132***	0.088**
	(0.020)	(0.051)	(0.038)	(0.038)	(0.034)
AVGVERT_INT	0.004*	0.031*	0.903***	0.040***	0.001
	(0.003)	(0.016)	(0.067)	(0.011)	(0.003)
AVGLEVERAGE	0.000	0.001**	0.000	0.000	0.000
	(0.000)	(0.000)	(0.000)	(0.000)	(0.000)
AVGDEBT_MAT	−0.112***	−0.198***	−0.071	−0.023	−0.153***
	(0.025)	(0.068)	(0.045)	(0.047)	(0.041)
HighGrowth $_{t0}$	−0.009**	0.015	0.033**	−0.007	−0.008
	(0.005)	(0.025)	(0.016)	(0.009)	(0.008)
Observations	42,117	5268	8772	10,536	17,541
Chi2	3319.1***	423.3***	563.1***	1036.7***	1395.7***
LogLik	−16,027.2	−1983.5	−3319.3	−3883.0	−6678.5

Notes: The dependent variable (HighGrowth$_t$) is equal to 1 in year t for firm i if the corresponding growth rate of TFP falls within the top 15% of the related sectoral distribution of the TFP growth rates. Marginal effects are reported. Significance levels: * 90% ** 95% *** 99%.

reported in the previous TPMs and indicate the presence of substantial persistence. Being in the top 15 per cent of the distribution of TFP growth rates in year $t - 1$ has a positive and largely significant impact on the likelihood of the firm still being in the top 15 per cent in year t, with a marginal effect at means of about 27 per cent.[10] What is relevant in the proposed analytical framework is that, even after accounting for firm-level time-varying factors, there is still a significant impact of the lagged dependent variable. This implies that the persistence detected in the descriptive analysis is not spurious.

Moreover, the results for the covariates provide interesting insights that can be used to qualify such persistence. First, the estimated relevance of contingent factors allows us to exclude the ergodic nature of the process under scrutiny. The evidence shows that the dynamics is influenced by the past; but it is not past dependent as it is sensitive to changes along the path. This confirms its path-dependent nature. Second, with respect to the analysis of firm-level control variables, the strategies pursued by companies appear to have a significant effect on persistence dynamics. In this respect, the negative and significant effect of vertical integration can easily be interpreted. Those companies that have reduced their vertical integration on average had a significantly higher likelihood of being among the best performers in terms of TFP growth rates. This evidence confirms that specialization strategies in high value added enhances the possibility of obtaining long-lasting outperformance in productivity growth.

The variable related to debt management highlights how those companies that have been able to finance long-term investments through credit channels show higher TFP growth rates. This may mean that this subsample of companies is less financially constrained and does not have to rely solely on internal cash flows to finance growth- and productivity-enhancing assets. The summary statistics from the sample in fact reveal that a significant share of companies have a very limited incidence of long-term debt, meaning that these companies implicitly use (or are forced to use) external financial sources with a maturity of less than a year to support assets that are defined in the long run. While in this analytical setting it is not possible to assess whether such apparently irrational behaviour is determined totally by external constraints (that is, inefficiency of the credit markets), the data still provide a clear indication of the non-trivial effects of the sources of finance. Finally, we do not find a significant effect associated with the variable capturing intangible intensity whose effects are probably absorbed by past TFP growth or limited by the broad definition of this variable in companies' financial accounts data.

With respect to other variables reflecting firm-level characteristics, as expected, size has a positive and significant effect, while we have identified a non-statistically significant effect of past levels of return on equity on subsequent TFP growth rates. This latter result might be due to the fact that, in the present sample, companies within an industry tend to differ more in terms of operational efficiency than ROE along time.

In order to take into account the effects of system properties, regressions were run for different sub-samples of companies. Columns II–V of Table 8.5 show the results for sub-samples referring to High-Tech and Low-Tech sectors during the expansion and contraction phases of the business cycle. The results confirm the descriptive evidence presented in the previous section on the relevance of system properties, and show differentiated dynamics of persistence across sectors and in time. In particular, after the year 2001, the magnitude of the coefficient associated with the lagged dependent variable significantly increases in both sector groups, suggesting that the macroeconomic cycle affects persistence dynamics in productivity growth. Moreover, sectoral systems appear to play an important role since, in more technology-intensive industries, persistence effects are always greater than in lower intensive ones.

Table 8.6 synthesizes all results obtained on the lagged dependent variable for all tests on different sub-samples including the regional split.[11] System properties are seen to be relevant in shaping persistence patterns also when the regional component is taken into account. There is in fact

Table 8.6   Marginal effects of lagged dependent variable (HighGrowth t − 1) for different samples based on the model specification reported in Table 8.5

Region	Sector	Period	HighGrowth $_{t-1}$ Marginal effect	Stand. error
ALL	ALL	1998–2005	0.269***	(0.007)
ALL	HITECH	2001–2005	0.309***	(0.013)
ALL	HITECH	1998–2000	0.232***	(0.047)
ALL	LOWTECH	2001–2005	0.271***	(0.010)
ALL	LOWTECH	1998–2000	0.112***	(0.030)
LOW R&D	HITECH	2001–2005	0.282***	(0.019)
LOW R&D	HITECH	1998–2000	0.247***	(0.037)
LOW R&D	LOWTECH	2001–2005	0.255***	(0.014)
LOW R&D	LOWTECH	1998–2000	0.106**	(0.042)
HIGH R&D	HITECH	2001–2005	0.327***	(0.018)
HIGH R&D	HITECH	1998–2000	0.188***	(0.066)
HIGH R&D	LOWTECH	2001–2005	0.286***	(0.014)
HIGH R&D	LOWTECH	1998–2000	0.126***	(0.046)

Note: Significance levels: ** 95% *** 99%.

huge variance in the magnitude of marginal effects of the lagged dependent variable, ranging from 0.327 for companies in High R&D regions, High-Tech sectors observed in the years 2001–2005 to 0.106 for companies in Low R&D regions, Low-Tech sectors observed in the years 1998–2000.

This result, interpreted in the context of the previous literature, sheds some light on how and why Italian economic systems may differ from those of other countries, in particular the US and the UK. The important role of external factors in supporting the persistence of productivity growth might be associated with the low levels of industrial concentration of the Italian economic system characterized by the pervasive role of small firms. Here system properties, and specifically knowledge externalities, play a much stronger role than in economic systems characterized by large corporations for which the persistence of productivity growth relies more on internal factors.[12]

Finally, we test the significance of size effects on the magnitude of the hysteretic influence on current TFP growth of past productivity performances by adding to the previous model specification an interaction term between the lagged dependent variable and a dummy variable which takes value 1 if the company has more than 250 employees (LARGE FIRM DUMMY). As indicated by results reported in Table 8.7, the interaction term is not significant in all the specifications. This seems to suggest that after controlling for all firm-level characteristics, including the direct effect on productivity dynamics exerted by company size, there is no productivity growth 'persistency premium' for large companies, since the magnitude of the effect of the lagged dependent variable does not significantly change across firms of different size. This result points out that persistency patterns can be independent of size, and that not only large corporations are capable of sustaining higher productivity performances in time.

5.   CONCLUSIONS

This chapter provides empirical evidence on the crucial role of the interplay between the internal characteristics of companies – including their size, management strategies and system properties (such as access conditions to local pools of knowledge and the dynamics of economic activity) – in assessing the path-dependent persistence of productivity growth. These results confirm the hypothesis that productivity persistence is an emergent system property that takes place when there is appropriate matching between the system properties and the characteristics of individual agents. The analysis of the persistence of total factor productivity (TFP) growth was conducted through the use of transition probability matrices

Table 8.7   Dynamic probit model on the persistence of TFP growth rates (model specification with interaction between lagged dependent variables and LARGE FIRM DUMMY)

Sector	1998–2000	1998–2000	2001–2005	2001–2005
	High-Tech	Low-Tech	High-Tech	Low-Tech
HighGrowth $_{t-1}$	0.224***	0.112***	0.304***	0.266***
	(0.047)	(0.030)	(0.014)	(0.010)
HighGrowth t − 1 * LARGE FIRM DUMMY	0.070	0.027	0.022	0.042
	(0.045)	(0.041)	(0.023)	(0.026)
SIZE $_{t-3}$	0.068***	0.081***	0.057***	0.083***
	(0.011)	(0.008)	(0.009)	(0.008)
ROE $_{t-3}$	−0.018***	−0.009***	0.001	0.000
	(0.005)	(0.002)	(0.001)	(0.000)
INTANG $_{t-3}$	−0.015	0.060*	−0.022	0.027
	(0.045)	(0.033)	(0.032)	(0.027)
VERT_INT $_{t-3}$	−0.190***	−1.008***	−0.317***	−0.299***
	(0.032)	(0.064)	(0.026)	(0.024)
LEVERAGE $_{t-3}$	−0.007	−0.032***	0.010	0.015
	(0.006)	(0.012)	(0.006)	(0.010)
DEBT_MAT $_{t-3}$	0.113**	0.066**	0.061*	0.104***
	(0.046)	(0.032)	(0.033)	(0.029)
AVGSIZE	−0.059***	−0.081***	−0.054***	−0.080***
	(0.011)	(0.008)	(0.009)	(0.008)
AVGROE	0.009**	−0.000	0.002	0.001
	(0.004)	(0.001)	(0.002)	(0.001)
AVGINTANG	0.038	−0.039	0.130***	0.084**
	(0.051)	(0.038)	(0.038)	(0.034)
AVGVERT_INT	0.031*	0.905***	0.040***	0.001
	(0.016)	(0.068)	(0.011)	(0.003)
AVGLEVERAGE	0.001**	0.000	0.000	0.000
	(0.000)	(0.000)	(0.000)	(0.000)
AVGDEBT_MAT	−0.188***	−0.070	−0.020	−0.149***
	(0.069)	(0.045)	(0.047)	(0.041)
HighGrowth $_{t0}$	0.015	0.033**	−0.007	−0.008
	(0.025)	(0.017)	(0.009)	(0.008)
Observations	5268	8772	10,536	17,541
Chi2	426.37***	565.90***	1037.47***	1398.71***
LogLik	−1981.9	−3319.01	−3882.71	−6677.0

(TPMs) which were split according to different system property dimensions, providing interesting results. The subsequent econometric analysis of firm-level TFP has shown that firms which have been able to improve the general efficiency of their production process at time t are more likely to sustain above-average performance in subsequent periods of time than firms with lower past rates of TFP growth. Moreover, our analysis identified persistency patterns that are independent of size, suggesting that not only large corporations are capable of sustaining higher productivity performances in time, hence confirming the significant role SMEs may have in enhancing productivity dynamics of economic systems.

The identified persistence turned out to be path dependent rather than past dependent since it is shaped by a number of complementary and contingent factors that affect locally the dynamics of the process. The identification of the path-dependent character of persistence in productivity growth helps us understand and appreciate the variety of results in the previous literature. The differences in the results of an increasing array of empirical investigations can be interpreted as follows: innovative activities have indeed potential hysteretic effects that only become actual persistence in productivity growth when a number of complementary and contingent factors concur to making the process actually non-ergodic. At each point in time, the creative reaction of firms and the probability of introducing, adopting and imitating further innovations and of outperforming competitors in TFP growth are in fact affected by the sequence of results in the past, but are also conditioned by the actual levels of their internal dynamic capabilities to accumulate and exploit technological knowledge and human capital. These in turn are influenced by the changing characteristics of the system in which firms are embedded, confirming that the persistence of productivity growth shares the intrinsic characteristics of an emergent system property. Considering that not only the dynamics but also the structural characteristics of economic systems are relevant in shaping persistency patterns, differences across countries in the way system properties may influence firms' productivity persistence might be relevant. This comparative aspect of the analysis can certainly represent an interesting issue to be scrutinized in further research.

The results of the investigation carried out through this chapter on the characteristics and determinants of the persistence of productivity growth contribute to the analysis of path dependency. A large body of literature has explored the path-dependent characteristics of the direction of dynamic processes. Relatively less attention has been paid to the path-dependent character of the rate of dynamic processes. The results of our analysis confirm that the persistence of productivity growth at the firm level is a path-dependent dynamic process from several viewpoints:

- The persistence is clearly non-ergodic because dynamic performances in the past affect performance over time.
- The rates of persistence are influenced by the past, but are also strongly sensitive to events that take place along the process.
- 'Small events' along the process may change not only its direction but also its rate.
- 'Small events' are internal to each firm, such as managerial practices, changes in routines and decision-making mechanisms.
- 'Small events' are also external, such as changes in access and use conditions of the stock of technological knowledge available in the system and in the networks of relations among firms that qualify them.

The distinction between path and past dependence is once more fertile: the persistence of productivity growth is indeed a path-dependent process but not at all a past-dependent, deterministic one.

NOTES

1. Our previous analyses studied persistence in innovation (Antonelli et al., 2012), whereas this chapter focuses specifically on the determinants of persistence in productivity growth, taking into account: i) the role of firm characteristics such as size, evolving internal capabilities and management strategies; and ii) system properties such as the macroeconomic, sectoral and regional contexts in which persistence displays its effects.
2. The level of yearly depreciation of physical capital was chosen following the approach in previous studies that applied perpetual inventory techniques to estimate yearly fixed capital levels, adopting depreciation parameters in the 5–10 per cent range for physical capital. Since the adopted depreciation parameter is constant across industries, changes should not be expected in the significance of estimate coefficients for slight changes in δ.
3. This measure of persistence is substantially different from the one adopted in Antonelli et al. (2013). As in the previous study, the state variable simply reflected the existence of positive changes in TFP over time.
4. Let P_{ij} and $\hat{P}_{ij}$ denote the population and sample probabilities of transition of a company from status i to status j. This transition can also be seen as the outcome of a binomial distribution. Hence, standard errors of the estimated transition probabilities can be calculated as a binomial standard deviation: $\sqrt{P_{ij} \times (1 - P_{ij})/N}$ where N equals the number of companies in status i. As N increases, $\hat{P}_{ij}$ tends to P_{ij}.
5. Sectors were divided into High-Tech and Low-Tech according to the Italian three-digit ATECO industry classification.
6. In order to simplify the description of results, we report only the probability associated with persistent top performers. Moreover, to be consistent with subsequent econometric models, controlling for the same system-level factors in productivity persistence, we consider the same selection criteria to identify top performers – that is, the first 15 per cent in the sectoral distribution of TFP growth rates. Persistency patterns do not significantly change with respect to different thresholds. All results are available upon request from the authors.
7. We carried out a sensitivity analysis to investigate whether, and to what extent, the

results are related to the selected threshold. Results are largely confirmed for different thresholds and are available upon request.

8. This intuition is particularly relevant given that the sample is composed of manufacturing companies that operate to a large extent in traditional sectors and that during the observed years have carried out significant restructuring outsourcing of production activities.

9. The empirical evidence on the relationship between profitability measures and productivity is mixed, even when taking into account operative profitability (ROI or ROA). In general, the identification of linear effects appears to be difficult. See Antonelli and Scellato (2011) for a discussion.

10. We also tested different specifications for TFP growth rate, using both a two-year and a four-year interval. Results were not affected.

11. Given that the effects of relevant control variables do not largely differ across estimates produced on different samples, for reasons of space we do not report results for all estimated coefficients. All results are available upon request from the authors.

12. See Antonelli et al. (2014) and Antonelli and Scellato (2015) for complementary evidence. This interpretation, for which we acknowledge the suggestion of an anonymous referee, might be the object of further empirical investigations on a comparative basis.

APPENDIX: ROBUSTNESS CHECKS

Table 8A.1   *TPM on quartiles of sectoral distribution TFP growth rates (all years and companies)*

	High Growth t	Mid–High Growth t	Mid–Low Growth t	Low Growth t
High Growth $t-1$	0.4748 (0.0048)	0.2496 (0.0042)	0.1617 (0.0035)	0.1134 (0.0030)
Mid–High Growth $t-1$	0.2472 (0.0041)	0.3343 (0.0045)	0.2585 (0.0042)	0.1595 (0.0035)
Mid–Low Growth $t-1$	0.1576 (0.0035)	0.2624 (0.0042)	0.3356 (0.0045)	0.2442 (0.0041)
Low Growth $t-1$	0.1169 (0.0031)	0.1563 (0.0035)	0.2471 (0.0042)	0.4795 (0.0048)

Note: Standard errors in parenthesis.

9. The endogenous dynamics of pecuniary knowledge externalities

Cristiano Antonelli and Gianluigi Ferraris

1. INTRODUCTION

This chapter contributes to the literature that impinges upon the approach elaborated by Schumpeter (1947), according to which innovation is the result of the creative reaction of firms facing unexpected changes in product and factor markets, contingent upon the availability of knowledge externalities. The availability of knowledge externalities, in turn, is the stochastic result of the introduction of innovations. Its persistence depends upon the actual amount of knowledge externalities that are generated at each point in time. This dynamics is the result of the interaction between individual decision-making embedded in a system and the changing conditions of the system (Antonelli, 2011, 2017a, b; Arthur, 2014).

The introduction of innovations requires the generation of technological knowledge. In turn, the generation and dissemination of technological knowledge can only take place in organized contexts characterized by appropriate levels of knowledge connectivity qualified in terms of viability of knowledge interactions and transactions among heterogeneous and creative agents that act intentionally to innovate when their individual performance is out-of-equilibrium. The generation of technological knowledge is, in fact, based on the interactive and collective recombination of internal and external knowledge through the intentional interaction and participation of a variety of learning agents embedded in geographic and professional knowledge commons. Interaction is required for the acquisition and implementation of external knowledge, an essential input in the generation of new knowledge (Antonelli and David, 2016).

This process leads to the generation of knowledge stemming from internal research activities combined with knowledge externalities and strategic mobility across knowledge commons. The outcomes are determined by the structured contexts in which they are embedded; but they are also the cause of changes in the structure of the system, its knowledge connectivity and the pecuniary knowledge externalities available within the knowledge commons, the likelihood of successful innovation and, ultimately, aggregate productivity. Innovation and changes to productivity levels affect the system's price levels and the performance of firms, promoting new out-of-equilibrium conditions and new structures of the system (Antonelli, 2008, 2011, 2015a, 2017a, b).

This open-ended feedback system is based on continual interactions between individual acts and endogenous knowledge externalities related to the structure of the system and its levels of knowledge connectivity. In this context, the decisions to both generate technological knowledge and introduce technological innovations by exploiting the knowledge interactions and organized structures in which they take place are endogenous and are determined internally by the dynamics of the system. The individual and intentional actions of creative agents are central to the system dynamics; however, no single agent is solely responsible for or is able to forecast the eventual results of his or her actions because of the effects on the organization of the system (Miller and Page, 2007).

The characteristics of the landscape in which knowledge interactions and transactions take place play a central role in assessing the viability of knowledge generation strategies. Thus, feasibility of knowledge generation depends on the knowledge connectivity of the system as measured by the levels of knowledge externalities, which, in turn, depend on the characteristics of the knowledge landscape. These characteristics are neither static nor exogenous. They change continuously through time as a consequence of the activities of agents and their ability to generate knowledge and introduce innovations and freedom to search for new opportunities for the generation of new technological knowledge. Changes to the features of the landscape engender both positive and negative externalities which affect the ability of firms to innovate. The changing capabilities of firms to generate new technological knowledge affect their mobility and, ultimately, the contours of the space. Moreover, knowledge landscapes and knowledge externalities are not given, but emanate from an endogenous, path-dependent collective process that includes institutional changes such as the introduction of new intellectual property right (IPR) regimes (Sorenson et al., 2006).

The present chapter draws on the above to build a synthetic account of the role of externalities in the economics of technological knowledge, implementing the notion of endogenous knowledge externalities, showing the dynamic endogeneity of the emergence and decline of knowledge externalities at the system level, and exploring their implications for the rates of introduction of innovations and productivity increases in the system. Section 2 reviews the changing attitudes to knowledge externalities, and elaborates a theoretical framework to understand the endogenous dynamics of pecuniary knowledge externalities. Section 3 presents an agent-based model of the innovation system. Section 4 presents the results of the simulation, focusing on alternative hypotheses related to the institutional and architectural features of the innovation system. Section 5 concludes by summarizing the main results and discussing some policy implications of the analysis.

2. KNOWLEDGE EXTERNALITIES AS INPUT AND OUTPUT OF SYSTEM DYNAMICS

Recent efforts to apply complex system analysis to the social sciences and to implement an economics of evolutionary complexity using agent-based simulation models (ABM) are particularly helpful to analyse the generation of technological knowledge as an endogenous collective process that is both the key causal factor and the outcome of system dynamics. In this approach, technological knowledge and innovation constitute the emergent property of organized contexts characterized by qualified interactions among heterogeneous and creative agents able to react intentionally to innovate when their performance is out-of-equilibrium. The individual and intentional actions of creative agents are central to the system's dynamics, which are determined by the structure of the system and the endogenous dynamics of knowledge externalities. No individual agent can claim responsibility for or forecast the eventual results of its actions. The complexity of the system is promoted by the interdependence between individual action and structural change (Lane, 2002; Lane et al., 2009; Page, 2011).

Following the knowledge recombinant approach, in order to generate new knowledge firms need to combine internal sources of knowledge, such as in-house research and development (R&D) activities and learning processes, with the systematic (as opposed to the occasional, additive) use of external knowledge, which is acknowledged to be an indispensable input for the production of new knowledge. Its criticality for the generation of recombinant knowledge to produce new technologies forces learning agents to search for and access it intentionally. No firm can innovate in isolation. External and internal knowledge sources are substitutes only to a limited extent: complete substitution between internal and external knowledge is impossible. External and internal knowledge, both tacit and codified, are complementary inputs – neither can be dispensed with (David, 1993; Weitzman, 1996; Fleming, 2001; Fleming and Sorenson, 2001; Cowan and Jonard, 2004, Antonelli and Colombelli, 2015a, b).

The limited appropriability of knowledge engenders flows of knowledge spillovers. Their actual absorption and eventual use in the generation of new technological knowledge, however, is determined by the knowledge connectivity of the system. In turn, the knowledge connectivity of the system is influenced by: i) the actions of learning agents that affect the structure of the system; ii) the knowledge interactions combined with internal learning efforts that affect the distribution of the knowledge possessed by each agent and made accessible through knowledge interactions. Similarly, mobility across the knowledge commons affects the density of agents and, hence, the amount of knowledge absorption costs.

Knowledge spillovers, in fact, do not automatically benefit all potential recipients (Griliches, 1979, 1992; Romer, 1990). Systematic and intentional efforts are required to exploit knowledge spillovers. This requires a knowledge exploration strategy to search, screen and identify knowledge sources, and to assess whether and to what extent the firm can rely on that source combined with the stock of internal knowledge to produce new knowledge. The firm must be able to fully combine and coordinate the relevant learning and research conducted within its boundaries with the relevant sources of tacit and codified external knowledge for the successful generation of new knowledge (Beaudry and Breschi, 2003; Bresnahan et al., 2001; Antonelli and Colombelli, 2015a, b).

Identifying and accessing external knowledge are expensive pursuits due to direct purchasing costs, whether there are markets for the knowledge and, especially, the costs of knowledge absorption. Knowledge interactions are required to access external knowledge – especially its tacit components – to reduce the risks to the vendor of opportunistic behaviour and knowledge leakage. It is difficult and costly to detail all the ingredients, necessary procedures, possible applications and implications of knowledge, and transfer of technological knowledge requires systematic codification efforts (Arrow, 1969; Mansfield et al., 1981; Lundvall, 1988).

Knowledge is sticky; it is embedded in organizations, protocols and procedures. External knowledge acquisition and sharing can be achieved only via direct and purposeful interactions to create the appropriate institutional context, which entail specific costs. The capacity of agents to access external technological knowledge depends on the fabric of the relevant institutional relations, and on shared codes of understanding that help reduce information asymmetries, limit the scope for opportunistic behaviour and build a context that allows reciprocity and the building of trust and generative relationships (Antonelli and David, 2016). The receptivity of firms to knowledge generated elsewhere is not obvious. Its absorption requires dedicated activities that have a cost and vary across firms (Cohen and Levinthal, 1990; Antonelli, 2011).

The use of external knowledge as an input in the generation of new knowledge entails knowledge absorption costs related to: i) knowledge transactions, communication and interaction costs associated with exploration activities such as search, screening, processing, contracting and interacting with competitors, suppliers and customers; and ii) the processing costs associated with the access and actual use of external knowledge (Griffith et al., 2003; Guiso and Schivardi, 2007). In some specific locations heavy knowledge absorption costs make access to external knowledge expensive. In others, knowledge absorption costs are low because of ease of access to the knowledge commons. These conditions are highly idiosyncratic and localized (Bischi et al., 2003; Zhang, 2003).

Pecuniary knowledge externalities are defined by the gap between the equilibrium cost of knowledge as an input in knowledge generation and its actual cost, taking into account its limited appropriability and exhaustibility.[1] Because of its limited appropriability knowledge cannot be fully appropriated and spills. Because of its limited exhaustibility it can be used again and again as an input in the generation of further knowledge. Its secondary use requires dedicated activities, and hence absorption costs. The costs of the secondary use of knowledge *may* be – in appropriate circumstances and favourable conditions of knowledge governance within economics systems – lower than the equilibrium levels of knowledge as a standard good. Pecuniary knowledge externalities are defined by the gap between the cost of knowledge as a standard good and the actual cost of knowledge, taking into account its limited appropriability as well as its absorption costs.

The levels of the pecuniary knowledge externalities available within the knowledge commons, the resulting amount of knowledge that the overall system can generate and the aggregate outcomes of the dynamics related to productivity levels are simultaneously endogenous and unpredictable, and subject to the changing interplay between individual action and structural change. In this approach, neither interactions nor the organized structures in which they take place are exogenous; they are determined internally by the system dynamics (Arthur et al., 1997; Lane et al., 2009; Antonelli, 2011).

The levels of pecuniary knowledge externalities vary across commons and time. They depend on the density of the co-localized innovation agents in the region. The density of knowledge commons yields, in fact, both positive and negative effects on the actual levels of knowledge absorption costs, and hence on the levels of pecuniary knowledge externalities. Density has negative effects on the amount of resources that are necessary to perform the exploration and search of external knowledge: the larger the density, the more expensive the identification of the external knowledge items that are necessary to generate new knowledge. Density, however, has also positive effects in terms of information processing. The larger the density, the lower the unit costs of the commons within which each firm is located. Total knowledge absorption costs, as a consequence, decline with density until a minimum is reached. Beyond a threshold level of density, where knowledge absorption costs hit a minimum, knowledge absorption costs increase along with the density of the commons. The relationship between density and net pecuniary knowledge externalities exhibits the typical traits of a U-shaped functional form.

At each point in time, the actions of agents, including the generation of new knowledge and the introduction of innovations, affect: the structure of the system; the architecture of networks; the density and quality of commons; the organization of communication flows; and, ultimately, the determinants of external knowledge availability and its governance costs. Specifically, the mobility of agents in the regional space, related to accessing external knowledge available within a rich knowledge commons, has a direct effect on location costs as well as on knowledge governance costs. Both too little and too much density of agents can be detrimental to the accumulation and creation of firms' technological knowledge and innovation capabilities. This refers to the notion of endogenous knowledge externalities.

The characteristics of the system into which knowledge flows matter in relation to knowledge governance costs, which include transaction, interaction, absorption and communication costs (Arrow, 1969). Because of the intrinsic non-exhaustibility and non-divisibility of knowledge, and its tacit and sticky characteristics, the cost of external knowledge may differ from the long-run equilibrium cost defined by matching marginal costs with marginal production. This important U-relation is strongly influenced by the level of knowledge governance costs that reflect the structure of the system. Only if the costs of external knowledge are below the equilibrium level will firms react by innovating. The introduction of innovation is clearly an emergent property of the system, which occurs only in specific and positive geographic, institutional and sectoral contexts. However, the structural characteristics that yield net positive knowledge externalities and the resulting introduction of technological innovations are local rather than global, are far from being static or exogenous, and are determined by strong endogenous and localized dynamics (Krugman, 1994).

As a result, net positive knowledge externalities are a transient property of the system in which firms are embedded. Schumpeter (1942: 28) commented that: 'Surplus values may be impossible in perfect equilibrium, but can be ever present because that equilibrium is never allowed to establish itself.' The quality of the knowledge governance mechanisms in place is important when assessing the size of the net positive effects of knowledge externalities.

Pecuniary knowledge externalities are endogenous to the system in reflecting the changing distribution of co-localized members of the knowledge commons. They are inherently path dependent in stemming from elements of past dependence demonstrated by the stock of firms in the knowledge commons at each point in time, through the pervasive role of contingent factors such as local interactions, feedback and strategic mobility of firms. The mobility of firms affects the net positive externalities available in each location. The entry of new firms is likely to increase the overall levels of knowledge governance costs and, at the same time, may increase the opportunities for knowledge sharing. On the other hand, firms' exit indeed helps reduce overall levels of knowledge governance costs, but also affects the opportunities for knowledge sharing. The mobility of firms is fully endogenous; it arises from the search for better opportunities to generate new technological knowledge, promoted by out-of-equilibrium conditions. At the same time, firms' mobility, by changing the structural conditions of the system and its knowledge connectivity, affects the actual opportunities for generating new technological knowledge.

The ruggedness of the system in which firms are localized is not an exogenous characteristic, as it is assumed in NK models; it is intrinsically endogenous and determined by firm mobility.[2] The dynamics of the system feeds continuously on the interplay between out-of-equilibrium conditions, firms' reactions, enhanced learning processes, external knowledge search, mobility in the knowledge space, structural changes, a new balance based on knowledge externalities, the generation of new technological knowledge, introduction of productivity-enhancing technological innovations, price reductions and eventual new out-of-equilibrium conditions. Endogenous knowledge externalities are at the heart of the innovation system.

At each point in time there may be several solutions, but each will be different in its standard characteristics of stability and replicability. Equilibrium points are erratic. Small shocks engendered by the mobility of firms seeking to absorb higher levels of external knowledge have major effects at both the aggregate and disaggregate levels, and may push the system far beyond any given values – although not back to levels experienced in a previous phase. The performance of individual agents, and of the system at large, depends on the distribution within the system of agents across the knowledge commons, their density and interactions, and their knowledge endowments. Each of these elements is interdependent, and each stems from the dynamics of constantly changing collective dynamics.

Path dependence, because of the roles of learning and interdependence, exerts powerful effects. The stock of available knowledge and the systems of knowledge communication in place at each point in time catch the effects of past dependence. However, small events can change the direction and affect the rates of these changes so as to alter the trajectories set at the origin of the process (David, 2007).

3.   AN ABM EXERCISE

3.1   The Building Blocks of the Simulation Model

ABM allows exploration of the workings of the interactions, transactions and feedback between individual actions and the system structure that make up the simple but articulated economic system outlined in the previous section. ABM provides a tool to grasp the dynamics of the complex interactions among agents, through and between the environment and the agents within it, that arise from the simulation (that is, the model computation), without the need for extensive and detailed descriptions of the dynamics investigated. This approach models, in a parsimonious and simple way, the intrinsic complexity of the knowledge interactions that are allowed to affect the structure of the environment in which they take place (Axtell, 2005; Terna, 2009).

The ABM implemented in this section operationalizes, through the interactions among a large number of objects representing the agents in the system, the functioning of a typical complex process characterized by: i) a key role of knowledge externalities; ii) augmentation by the Schumpeterian notion of creative reaction conditional on the availability of knowledge externalities (Schumpeter, 1947; Antonelli, 2017a, b); and iii) enrichment by the explicit assumption that the actions of agents affect the structure of the environment, including the amounts of pecuniary knowledge externalities (Lane, 2002; Lane et al., 2009).[3]

The model assumes bounded rationality of firms, and is based on appropriate criteria of conduct related to procedural rationality. Firms are endowed with the abilities to learn and react that enable procedural rationality augmented by the inclusion of potential creative reactivity. Firms are credited with the ability to try to react: their reactions are determined by the out-of-equilibrium conditions when profitability levels are far from the average. Their reactions are creative and, when and if positive knowledge externalities are available, lead to the introduction of productivity-enhancing innovations rather than only adaptations or adjustments between quantities and prices (Antonelli, 2008, 2011).

In the ABM, demand and supply meet in the market place, production is decided *ex ante* and firms try to sell their output in the product market, where customers spend their revenue. The matching of demand and supply sets temporary prices that define the performance of firms. Firms are heterogeneous both with respect to their productivity levels and ultimate profitability, and with respect to their location. The economic system is represented as a collection of regions, or commons, across which firms are distributed at the start of the simulation process.

In the simulation, heterogeneous firms produce homogeneous products that are sold into a single market. In the product market, households expend the revenue derived from wages (including research fees) and the net profits of shareholders. In input markets, the derived demand from firms matches the supply of labour provided by workers, including researchers. For simplicity, no financial institutions are activated, and payments cannot be postponed. Firms' capital is supplied solely by shareholders, and all the commercial transactions are cleared immediately. Market clearing mechanisms based exclusively on prices maintain a perfect equilibrium between demand and supply. This equilibrium is ensured for both product and factor markets: quantities determine the correct price, enabling the whole production to be sold. No friction or waiting times are simulated; factors are assumed to be immediately available.

The production function is very simple and avoids issues related to different kinds of production processes, input availability, warehouse cycles and so on: outputs depend exclusively on the amount of employed labour and its productivity. Both labour and productivity vary among firms. Labour depends on the entrepreneur's decision about the growth of production; productivity is a function of the technological level achieved by the firm via innovation.

The whole output is sold in the single product market, where the revenue equals the sum of wages, dividends and research expenses, and the price depends on liquidity. According to the temporary price levels, profits are computed as the difference between income and costs, no taxes are paid and no part of the profit is retained by the enterprise. Shareholders either receive profits or reintegrate losses. Firms can support their losses only up to a certain threshold beyond which they leave the market and are replaced by new entries, after a parametric number of production cycles.

Firms are learning agents that are able to react to out-of-equilibrium conditions. According to their performance levels and the availability of external knowledge, firms can fund research activities dedicated to innovation. Firms learn internally by doing, and externally by interacting. Internal learning processes are intrinsic to the firm and occur spontaneously over time. External learning involves two aspects. First, the rate of internal learning is influenced by the local conditions of the commons. The accumulation of competences via the firm's learning processes is greater the greater the average productivity of all the other competitors co-localized in the commons.

Second, we assume that localization in a knowledge commons provides the opportunity to absorb technological knowledge from co-localized firms with higher levels of productivity. External learning entails specific knowledge governance costs required to carry out the necessary activities of knowledge networking and communication among all the members of the commons. Knowledge governance costs depend on the number of firms within each commons by means of both fixed and variable costs. Fixed costs stem from the administration of the commons: the level increases with the size of the commons, but unit costs for each firm decline as the fixed costs are shared with the other members of the commons, independently of the need and opportunity for external learning. Next to fixed costs there is the variable part of the knowledge absorption cost that is proportional to the number of firms in the commons. In this way the cost function that relates the amount each firm has to bear to participate in and take advantage of a commons, to the population of the commons, becomes a U-shaped curve.

The whole system is represented as a nested collection of agents; they are grouped in commons that are constituted by a simple collection of agents; the collection of commons constitutes the whole system (a collection of collections of agents). The simulation process shows that the localization of the agents in different commons is the result of their past activities, although these can change at each point in time. The results from a production and consumption cycle influence the strategies adopted by the agents during the next cycle. Hence, the dynamics of the model is typically characterized by path dependence: the dynamics is non-ergodic because history matters, and irreversibility limits and qualifies the alternative options at each point in time. However, at each point in time, the effects of the initial conditions may be balanced by occasional events that could alter the 'path', that is, the direction and the pace of the dynamics (David, 2007).

Firms perform basic search functions and acquire information about the levels of profitability of neighbouring firms in the same commons. As a result of bounded rationality, the firms in the model are not able to observe the entire economic system, but only the average levels of profitability of the other firms. Individual transparency is clearly local: the spectrum within which firms can observe the conduct of other firms is limited to the particular commons.

The further profitability lies outside the local average, the stronger the out-of-equilibrium conditions. If profitability results are below average, firms can innovate in order to improve their performance; when results are above average, they can take advantage of abundant liquidity and reduce the opportunity costs of risky undertakings. Innovation is viewed as the possible result of intentional decision-making that takes place in out-of-equilibrium conditions. The further the firm from equilibrium, the more likely it will innovate. Hence, we assume a U-shaped relationship between levels of profitability and innovative activity, measured by rates of increase of total factor productivity (TFP).

To summarize, the firm's motivation to innovate increases each time its performance is found to be far enough from the local average. The motivation becomes progressively stronger if the enterprise's relative position remains outside the band for several and consecutive production cycles: after a parametrically set number of consecutive cycles the enterprise performs an innovation trial.

Out-of-equilibrium conditions push firms to try to react by generating technological innovations that will increase their productivity. Attempts to generate new technological knowledge and to innovate are based on internal research and learning efforts, and access to external knowledge available within and across commons. Search for and access to external knowledge can be both local and global. When the neighbourhood in which each firm is embedded does not provide sufficient opportunities to generate additional technological knowledge, firms can move within knowledge space across commons to get closer to firms with high levels of technological knowledge. The absorption of external knowledge requires dedicated resources and specific costs, as does mobility across commons to achieve proximity to firms with higher levels of productivity.

Building on the growing empirical evidence on the intrinsic characteristics of agents' dynamics, we characterize the search activities at the base of the innovation process in our learning firms as typically displaying *Lévy flight* traits. We suppose that firms alternate extended phases of local search within their own commons with long jumps that take them to other commons (Barabasi, 2010). Hence, we assume that the generation of additional technological knowledge takes place when the learning firm is able to master a three-step sequence consisting of: i) valorization of internal competence based on learning processes; ii) local (within commons) absorption of external knowledge; and iii) entry into a new commons characterized by higher levels of net pecuniary knowledge externalities.

The successful generation of new technological knowledge at the same time yields new knowledge externalities and enables the introduction of productivity-enhancing innovations. Their introduction, in turn, reduces the overall price levels in the product markets (affects the working of factor markets) and creates new out-of-equilibrium conditions. The micro–macro dynamics loop is closed, and engenders continuous growth and change provided that changes to the system structure do not promote provision of positive net knowledge externalities. The interaction between individual action and systemic change includes the new knowledge externalities that spill from the limited appropriability of the new knowledge and the structural changes determined by the mobility of firms across the knowledge commons, and its effects on knowledge governance costs. Endogenous knowledge externalities are the engine of system dynamics. Their level is not given and static: it can increase and decrease according to the amount of innovation being introduced at each point in time, and hence the amount of knowledge generated at each point in time, taking into account the changing levels of knowledge connectivity determined at each point in time by the changing structural landscape of the system (Anderson et al., 1988; Rosser, 2004).

3.2   A Detailed Presentation of the Innovation Process Simulation

Since the chapter aims to identify the changing role of endogenous knowledge externalities in the innovation process, here we explore the ABM of the innovation process in detail, and stress analytically the role of the external factors that shape the recombinant generation of technological knowledge. Firms are characterized as learning agents. Learning is both internal and external to the firm:

- Internal learning is a routine that includes typical processes of learning by doing and learning by using. Internal learning enables the accumulation of tacit knowledge and potentially competence that requires a specific action to be eventually mobilized and transformed into concrete technological knowledge.
- External learning influences the rates of accumulation of each firm. It is also a routine and consists of monitoring activity that enables firms to assess the profitability and productivity of the other firms co-localized within the commons. External learning relies on interactions with other firms in the same commons. Bounded rationality confines firms to observing only other firms in their particular commons. External learning provides information on the availability of external knowledge that can be tapped if and when the firm tries to upgrade its productivity level. External learning encompasses two processes: i) faster learning rates, influenced by the average productivity of the commons; and ii) the possibility to absorb technological knowledge from co-localized firms with higher productivity levels.

Agents follow a satisficing approach in their decision to try to innovate. At each point in time, learning firms assess their own profitability against that of co-localized firms within the commons. If their profitability is either below or above the local average, the firm will react. Their reaction may be adaptive or creative according to the availability of knowledge at a cost that is below the marginal product: innovation efforts are expensive because innovation is not free. Firms are short-sighted and can expend, in one unit of time, all their innovation budget (including absorption costs) even when the productivity gains obtained from absorption extend over more than one unit of time. Innovation efforts can fail if the innovation costs exceed the productivity gains. In this case the reaction of agents will be adaptive. This takes place when the knowledge connectivity of the system is small and the levels of knowledge externalities are low. If knowledge is available at costs below its marginal product, the innovation efforts may be successful, resulting in a creative reaction.

The innovation process consists of three sequential phases. In the first, firms try to mobilize their internal slack competence. In the second, firms with insufficient potential competence based on past learning processes will try to absorb external technological knowledge spillovers from neighbouring firms in their own commons. If this is not possible, the third phase consists of a random move to another location in a different commons. Let us consider each of these in turn.

- Firms consider the possibility to change their production technology when their performance is out-of-equilibrium and differs from the average. Out-of-equilibrium conditions are the result of mismatches between expected and actual product and input market conditions. Firms in out-of-equilibrium conditions try to innovate. To innovate firms mobilize internal slack competence accumulated through learning processes and access to external knowledge. The firms in our model are endowed with the ability to improve their production cycles. With each production cycle, the firm acquires and cumulates some technological potential. This potential requires intentional and dedicated research activities for its transformation into innovation. Competence can be transformed into innovation at a cost. Internal slack competence however is not sufficient to support the recombinant generation of new technological knowledge and the introduction of a productivity-enhancing innovation: external knowledge is an indispensable, complementary input. In order to access and use external knowledge firms will try to access and absorb knowledge spilling from other firms. The search for external knowledge takes place locally within their own commons and at distance in neighbouring commons.
- Local absorption enables exploitation of technology introduced by other firms. Firms can take advantage of their information

acquisition from external learning processes, and can identify more profitable, co-localized firms. Absorption requires dedicated activities; and, due to absorption costs, it is not free. Effective access to external technological knowledge requires substantial resources for exploration, identification, decodification and integration into the internal knowledge base. The absorption of knowledge from firms with higher levels of productivity is neither free nor unlimited. First, absorption of external knowledge requires specific activities and resources that have a cost. The level of these costs depends on the productivity gap between knowledge recipient and possessor. Second, the knowledge connectivity of the system plays a major role. When knowledge absorption gives poor or null results, firms move to another location in order to better address their technological conditions.

- The third way to improve productivity levels involves moving around the physical space in order to identify more interesting commons (mobility across commons). When mobilization of competences and within-commons knowledge absorption are not viable solutions, firms can try to move randomly to another location in the hope of finding superior knowledge and a higher stochastic possibility to absorb technological knowledge from firms with high productivity levels. Since firms have access to individual information only about firms in their own commons and not all the other firms in the system, the Lévy flight is blind. This random move can lead to superior as well as inferior commons. Thus, firms decide to move only if the profitability of their commons is below the system average. If it is above the average, the chances of finding a superior commons will be low. The conduct of firms shapes the structure of the system and, at the same time, the structure of the system influences the innovation chances of firms in several ways. Localization in an advanced commons is beneficial because learning is faster and prospective recipients have higher possibilities to observe and absorb technological knowledge that high-productivity firms cannot fully appropriate. However, at the same time, localization in a dense commons engenders high costs of search and interaction, with the possible reduction of net pecuniary knowledge externalities.

3.3 Analytical Representation of the Simulation Model

This section presents the analytical organization of the simulation model and the founding equations.[4] The production activity is specified following a simple linear function:

$$O_i = A_i Lp_i. \qquad (9.1)$$

Where the output (O) of a generic i-th enterprise depends on the labour employed in the production cycle (Lp) and its productivity (A). The latter can vary between 0 and $+\infty$. Customers (that is, workers, shareholders and researchers) spend the whole amount they earn in buying goods, so the selling price for goods is simply computed as:

$$p = Y/\sum O_i. \qquad (9.2)$$

where Y represents the whole amount earned by the customers and the sum computes the total production of enterprises operating in the simulated economy. The amount of wages represents the full costs of the enterprises: research costs, as well as moving ones and costs related to the exploration of the commons (that depend on the size of the common), are simply computed as work units to be bought. The number of work units the enterprises demand for each cycle is determined as:

$$L_i = Lp_i + CC_i + T_i + M_i. \qquad (9.3)$$

where T_i represents the work units required to transform accumulated knowledge in technological innovation (either for internal learning or spillover from other firms into the commons); CC_i measures the work units needed to access the common knowledge base, including the research costs to increase the technological level by means of the external knowledge spilling in the commons where each firm is located (external learning); and M_i represents the work units needed for mobility across commons. Note that Lp_i represents a firm's entire input; in this way the whole stylized economy becomes quite simple.

The unit wage (w) for a single work unit is the same for each enterprise; it is centrally computed as a constant value equal to one, assuming an unlimited supply of labour:

$$w = 1 \qquad (9.4)$$

Each firm pays its workers a total amount of wages (W) of:

$$W_i = wLp_i. \qquad (9.5)$$

The total amount of wages is simply computable as:

$$W = \sum W_i. \qquad (9.6)$$

Firms decide to try to change their technology when their performance differs from the average in both cases of profits or losses. The resources invested to try to change their technology are defined in the former case by the amount of extra profit (this will be the maximum affordable investment), whereas in the latter such amount is measured by the savings the enterprise realizes by reducing its input (labour) acquisition. In this way the adaptive response of enterprises is driven by profits: with a loss they reduce the amount of factors demanded (and vice versa when they enjoy profits), whereas the reactive response is driven by the difference between the results of each single firm and the average results of the firms in the shared commons. The amount a firm invests, in case of internal learning or spillover, is computed as:

$$I_{i1} = \min (T_{i0}, \text{profit}_{i0}) \mid \text{profit}_{i0} > \text{tolerance} \qquad (9.7)$$

or as:

$$I_{i1} = \min (T_{i0}, (- \Delta L_{i1} * \text{labour price})) \mid \text{profit}_{i0} < - \text{tolerance} \qquad (9.8)$$

For enterprises that perform moving strategies equations (9.7) and (9.8) work as well by simply substituting M_{i0} with T_{i0}.

Note that one action only can be taken in each cycle. Firms invest their resources in three ways: i) to transform their accumulated competence and access external knowledge; ii) to transform spilled over technologies obtained by exploiting the information retrieved by belonging to a common organization; iii) to move to another commons in the hope – the flight is blind – of finding better conditions. Let us analyse these in detail.

The transformation of accumulated knowledge in new technology (so-called internal learning) can be performed only if a firm has accumulated a minimum amount of knowledge specified through the parameter 'productivityUpgrade' and could be performed for every amount greater than this amount at a time. The cost of the process is fixed in value, in work units, specified for the parameter 'transformationCost':

$$T_i = \text{transformationCost} * \text{internalLearning} / \text{productivityUpgrade} \mid$$
internalLearning > productivityUpgrade (9.9)

To access external knowledge, firms search in the knowledge commons and bear the knowledge absorption costs (CC_i) that are related to the size of the commons and included in the costs each enterprise has to pay to be part of it. The relationship between density and knowledge absorption costs is U-shaped. For low levels of density, the larger the density of the common, the lower the knowledge absorption costs. Beyond a threshold, after the minimum, knowledge absorption costs are larger the larger the density: the costs incurred to access and process information about the knowledge spilling from the other firms increase with density. These types of costs are computed, each cycle (in work units) and have the same value for each enterprise.

The knowledge absorption costs (CC) are parametrically determined in each simulation through the parameter 'commonCost', and depends on the number of firms in the commons (N). First, the fixed component is computed as commonCost times the theoretical maximum number of components – that is, the number of agents in the whole economy (N); this is spread evenly among the firms belonging to each commons (n_i). The variable part of the knowledge absorption cost is proportional to the effective number of firms that belong to the common (n_i). The following formula summarizes the costs each component of the i-th common has to bear to belong to it:

$$\text{cost}_i = (\text{commonCost} * N) / n_i + \text{commonCost} * n_i \qquad (9.10)$$

The amount of external knowledge each firm can access depends on its distance from the 'spilling' firm. This distance (delta) is computed as follows:

$$\text{delta} = (A_j - A_i) / A_j \qquad (9.11)$$

where i is the enterprise trying to access external knowledge spilling from firm j. Note that it is possible to take advantage only of technologies whose patent licence has expired. In order to transform spilled technologies the firm has to bear the transformation cost, in the same measure it has to bear to transform internal learning. To tackle the spillover each firm has to invest an amount in working unit that is:

$$T_i = \text{transformationCost} * (A_j - A_i) / \text{productivityUpgrade} \qquad (9.12)$$

Actual access to external knowledge takes place with a probability defined as spilloverMinProb. The probability the spillover was successful (Ss) is:

$$Ss = (1 - \text{spilloverMinProb}) * (1 - \text{delta}) + \text{spilloverMinProb} \qquad (9.13)$$

If no local knowledge pecuniary externalities are available either because spillover is not allowed or no firm provides suitable spillover in the commons (including the special case when the firm is unique in the commons), firms try to move to another commons. This activity has a fixed cost set to the value (in units of work) specified by the parameter 'movingCost':

$$M_i = \text{movingCost} \qquad\qquad (9.14)$$

The outcome of their move will be positive so as to fuel a creative reaction and introduce innovations when and if the cost of knowledge – after taking into account knowledge absorption costs and moving costs – is below equilibrium levels. The dynamics of the system is now fully set. Firms caught in out-of-equilibrium conditions, with performances that are below or above average, try to react by means of the introduction of innovations. In order to introduce innovations they try to take advantage of pecuniary knowledge externalities. To do so they may move from one knowledge commons to another. Their entry and exit affect the amount of pecuniary knowledge externalities available in each commons.[5]

3.4   The System Dynamics of Endogenous Knowledge Externalities

Let us summarize the key points of the ABM to stress the relevance of endogenous knowledge externalities for the system dynamics. Appreciation of the endogeneity of knowledge externalities captures the characteristics of endogenous growth shaped by the intrinsic path-dependent dynamics of the system at both the structural and macroeconomic levels.

At the start of the simulation, heterogeneous firms, localized in different commons, are endowed with different levels of productivity that are randomly distributed in the range]0,0.25[following a uniform probability distribution. Firms start the production process at their particular productivity level, try to sell their goods on the product market, and experience different levels of profitability. They compare their profitability with the average in the commons to which they belong. If their profitability is either below or above the local average in their commons, these firms will try to change their knowledge base and introduce technological innovations. These innovation efforts are deemed successful if their costs are below the value of their gains in terms of productivity in one unit of time. The costs of knowledge have a major influence on assessing the viability of innovation efforts.

Innovation efforts consist of a sequence that starts with the valorization of their internal competences based on internal learning processes influenced by local average productivity levels. If the internal competence is not sufficient to introduce a new technology in order to increase productivity, firms move on to the second step and build on the information gathered through knowledge governance activities to try to absorb knowledge from co-localized firms (within the same commons) with higher profitability. If no such firms exist locally, then they move to the third step and attempt to move out of the original commons. Bounded rationality prevents assessment of whether the level of the knowledge governance costs in the new commons is lower than the advantages stemming from the external knowledge. The leap is blind. In the case of a negative outcome, the firm will continue to move across the system, to other commons.

This mobility of firms has important consequences for the system's structural landscape and the endogenous generation of knowledge externalities. Location in a knowledge commons is expensive due to the knowledge governance costs entailed in the resources required for searching, screening and assessing the levels of knowledge of the neighbours, and the costs involved in activating communication channels and networking interactions with them. The density of firms in a knowledge commons determines the level of knowledge governance costs, with the result that the mobility of firms across commons affects the knowledge governance costs of all other commons members. Firm exits impact on knowledge governance costs too. Entry and exit impact may be either positive or negative depending on the number of firms belonging to the commons, that is, to the current position of the commons on the U-shaped cost curve. The levels of net pecuniary knowledge externalities available in a knowledge commons are strictly endogenous to the local system, with important dynamic effects.

The distribution in space of agents, scattered randomly at the beginning of the process, becomes fully endogenous as agents move across knowledge commons in the regional space in the search for access to external knowledge from the spillovers of proximate high-productivity firms. At the same time, since pecuniary knowledge externalities are endogenous, the actual level of net positive pecuniary knowledge externalities available at each point in time, within each knowledge commons, changes over time as a consequence of the mobility of learning agents and the consequences – in terms of knowledge governance costs – for all the members of the knowledge commons.

Hence, the dynamics of the regional distribution of agents exhibits traits typical of path dependence. The process is non-ergodic, but not past dependent: small variations may exert important effects in terms of emergence of a strong commons, or determine its decline and force firms to exit with their progressive dissemination in space. At the system level, excess entry in a 'fertile' knowledge commons may halt the generation of new technological knowledge and affect the rate of increase of productivity: excess knowledge governance costs reduce net positive pecuniary knowledge externalities to zero. This is most likely in commons populated by high-productivity firms since their higher levels of technological knowledge are likely to benefit firms that are willing to innovate and, having casually landed in such commons, will enjoy the possibility to exploit their new position.

The introduction of productivity-enhancing innovations affects the position of the supply curve and modifies the conditions of the product markets: prices as well as the profitability of all incumbents will fall. Firms will reassess their profitability levels with respect to the local average, and the process will keep going provided that changes to the structural conditions of the system promoted by the mobility of firms in the space have not engendered the provision of knowledge externalities. The mobility of firms is the prime internal factor in the endogenous dynamics of the landscape and, hence, in the endogenous determination of the levels of knowledge externalities that shape the viability of the innovation process at firm level (Antonelli, 2011). This loop affects the system in four ways. Specifically we expect to see the following:

- At the firm level, the levels of endogenous knowledge externalities may inhibit or foster the successful introduction of innovation.
- At the structural level, the dynamics exerted by the interplay between centrifugal and centripetal forces changes the structure of the system and the attractiveness of different commons. When knowledge governance costs exceed the benefits from external knowledge, centrifugal forces are at work: the density of commons declines with the exit of firms. Centripetal forces are at work when the benefits of external knowledge are greater than the sum of the knowledge governance costs: the size and density of the commons increases. The structure of the system is characterized by changing heterogeneous 'stains', indicating commons where the introduction of productivity-enhancing innovations takes place and commons where no innovation is possible. The distribution of these 'stains' changes continuously over time.
- At the commons level, the dynamics of output and productivity is characterized by typical Schumpeterian waves as the changing interplay between centrifugal and centripetal forces engenders different phases that affect the overall, aggregate rates of productivity and output growth which exhibit both growth and decline.
- At the macro-system level, the dynamics of the system is likely to exhibit a step-wise process of output and productivity growth. The wave-like change at commons level in aggregate engenders a positive outcome, with phases of fast growth shaped by the upsides determined by the prevalence of centripetal forces, and phases of slow growth where the downsides are due to the stronger impact of centrifugal forces.

4. RESULTS

The results of the simulation confirm that the model is consistent, and is able to mimic the workings of a complex system based upon a large number of heterogeneous agents – both on the demand and the supply side – that are price takers in product markets where they are able to make efforts to react to changing market conditions. Replication of the temporary equilibrium price in the long term confirms that the model is appropriate to explore the general features of the system when the reaction of firms is adaptive and consists only of price to quantity adjustments. In the extreme case where firms cannot innovate due to lack of internal competence to mobilize and lack of external knowledge to be absorbed, the system effectively mimics a static general equilibrium in conditions of allocative and productive efficiency, with no dynamic efficiency. The markets sort out the worst-performing firms and drive prices down to the minimum production costs. This result is important because it confirms static general equilibrium as a simple and elementary form of complexity that emerges when agents are unable to innovate. As soon as positive levels of knowledge externalities allow agents to react successfully to changing market conditions, by innovating, the equilibrium conditions turn dynamic and key system elements (such as price, quantities, efficiency and structure) keep changing (Antonelli, 2011, 2017a, b). The dynamics, however, is not steady: the action of firms may engender negative effects on the knowledge connectivity of the system that, in turn, reduces the levels of net pecuniary knowledge externalities.

The results of the simulations of the model confirm the crucial role of endogenous knowledge externalities: with no positive externalities, productivity growth is much lower compared to when externalities are at work. The dynamics of the simulated system exhibit a wave-shaped trend describing firms' continuous search for more profitable commons. These results were achieved using a plausible but not fully calibrated parameter configuration and, thus, need to be confirmed by a deeper investigation.

The simulation results confirm the existence of different areas within an economic system, where productivity grows at different rates and profits follow different distributions over time as an outcome of the endogenous effects of each firm's relocation decision. In this process, commons are continuously augmented and reduced: new firms arrive and existing firms move to other commons, with the balance between incoming and leaving agents mostly unable to maintain the commons' population stable. Thus, their size varies with each simulation step.

Depending on the capability of commons to retain agents, a single commons could operate as an attractor, dramatically expanding its size. As already mentioned, since the Lévy flight is blind, agents move randomly to a new commons, but do not move if their profits are close to the average profit at the macro commons level or their commons profitability is greater than the average profitability of the whole economy. The more a commons grows, the more the knowledge governance costs for firms increase. When the costs overcome the benefits due to net positive knowledge externalities, profits start to fall, inducing firms to relocate to try and find more profitable commons.

Simulations demonstrate that the distribution of firms and, consequently, the actual levels of net positive knowledge externalities are the product of an endogenous process. Starting from a uniform distribution of firms across ten commons, the continuous relocation of agents produces a sequence of growth and decay of the commons according to the level of net positive pecuniary knowledge externalities their aggregation is able to engender.

The high technological and productivity levels achieved by more-developed commons tend to become diffused as firms in these commons decide to move to less-developed locations. Average productivity levels are very similar among commons because, in less-developed commons, the higher knowledge brought by new entries from more-developed commons rapidly spills over due to centrifugal forces. The decay of a former extensive commons is the means of sharing the effect of knowledge externalities with other commons, and provides valuable opportunities for less-developed firms to make the leap towards higher productivity.

Specific simulations have been done to focus a number of key issues, such as the existence and effectiveness of positive externalities. The findings come from comparing the results for four scenarios differing in the intensity of externalities: i) Alpha represents the benchmark scenario, with full deployment of both types of knowledge externalities: internal learning enhanced by the average productivity of the commons, and opportunities to absorb external knowledge at low knowledge governance costs; ii) Beta excludes knowledge governance costs and enhanced internal learning, but includes the cheap absorption of external knowledge; (iii) Gamma excludes knowledge governance costs, but includes internal learning at a fixed rate based on accumulation of experience, and independent of the average productivity of the commons; (iv) Iota excludes knowledge governance costs and allows only internal learning at a fixed rate based on accumulated experience.

1. Alternative dynamics with respect to the benchmark scenario, Alpha, are where the accumulation of experience proceeds at a faster pace in more developed commons but knowledge governance costs increase more proportionally than population.

2.   Alternative dynamics with different numbers of commons (Theta scenario).

In order to enable full comparability of the results, all the simulations in the second group were computed using very similar parameter set-ups (few values change among the different scenarios), the same number of agents, same duration, same number of commons and same random distribution. Specifically, each scenario simulation was run for 2000 production cycles involving 1000 agents. Scenarios Alpha, Beta, Gamma and Iota used ten commons, while in scenario Theta agents are grouped in only four commons because this scenario studies the influence of a different dispersion of agents.

At the onset of the simulation, levels of productivity are scattered randomly for each firm between 0 and 0.25, following a uniform random distribution; firms are endowed with initial accumulated knowledge randomly distributed between zero and 0.1 – the minimum knowledge level that can be transformed into increased productivity.

Information flows among agents are allowed only within each commons, where agents are able potentially to observe at each moment all the other agents in that commons even when their number becomes quite large. Agents have no information on other commons, but do know the average profitability of the whole economy (macro-system level) and that of the commons they belong to (macro-commons level).

When an agent's cumulated losses exceed a parametrically fixed threshold, the agent exits the market and goes out of business. After a few cycles (another parameter) it is replaced by another agent endowed with technology equal to the average level in the commons. In order to exclude results that were simply due to random events, simulations sub 1 and 3 were run 100 times by varying the random seed – used to set up pseudo-random distributions – and their results are presented as average figures of the 100 runs, confirmed by the low level of the related variance.

4.1   Existence and Effectiveness of Externalities

As outlined above, the investigation compares the results obtained from running simulations of four scenarios (Alpha, Beta, Gamma and Iota), based on varying values of several key parameters – knowledge governance costs, the negative effects of knowledge appropriability on the price of innovated goods, and external opportunities – which influence the effects of localization in a commons on the accumulation of competence and the capability to absorb external knowledge.

In more detail, knowledge governance costs are computed for each

Table 9.1   Alpha versus others: set-up of the different scenarios

Scenario	Number of commons	Common cost	Internal learning	External learning
Alpha	10	(0.01*N)/n + 0.01*n	0.001*(1 + cp)	Yes
Beta	10	zero	zero	Yes
Gamma	10	zero	0.001	Yes
Iota	10	zero	0.001	No
Theta	4	(0.01*N)/n + 0.01*n	0.001*(1 + cp)	Yes

firm according to the density of the commons to which they belong. Density exerts a non-linear effect so that knowledge governance costs vary according to the number of firms belonging to each commons following a U-shaped relation. In the first scenario (Alpha) this dynamic is fully at work, whereas in the other three (Beta, Gamma and Iota) the knowledge governance cost is set to zero. The external opportunity parameter measures the effects of productivity external to each agent, which adds to each agent's internal knowledge stock, at each production cycle.

According to our model, firms localized in a high-productivity commons accumulate more competence than firms in a low-productivity one. This parameter takes three values: i) in the Alpha and Theta scenarios it is set to 0.001 times the average productivity of the agents in the commons plus one; ii) in the Beta scenario the experience accumulated in each production cycle is set to zero, that is, there is no cumulated experience; iii) in the Gamma and Iota scenarios, which mainly test the effectiveness of different set-ups for this parameter, the firms accumulate 0.001 of experience for whatever productivity levels are achieved in the commons. Table 9.1 presents the experimental set-ups, where N is the total number of enterprises in the economy; n is the number of enterprises belonging to a single common; and cp is the average productivity of all the firms belonging to a commons.

As Table 9.1 shows, knowledge governance costs are set to zero in the Beta, Gamma and Iota scenarios, and in the Alpha scenario are allowed to vary according to the magnitude of each commons' population by following a U-shaped relation. In the Alpha scenario firms achieve a larger accumulation of competence that reflects both the average productivity of the commons in which they are localized and its productivity peaks. However, in the Alpha scenario firms are liable for knowledge governance costs that vary according to the density of the commons (Table 9.2).

The main simulation result is based on a comparison of productivity growth across the three sets of parameters. We expect the Alpha scenario to exhibit the best performance. The interpretation of the results is straight-

Table 9.2   Alpha versus others: population and knowledge governance costs

Scenario	Population of the common							
	1	50	100	150	250	500	750	1000
Alpha	10.01	0.70	1.10	1.57	2.54	5.02	7.51	10.01
Beta	zero	zero	zero	zero	zero	zero	zero	zero
Gamma	zero	zero	zero	zero	zero	zero	zero	zero
Iota	zero	zero	zero	zero	zero	zero	zero	zero
Theta	10.01	0.70	1.10	1.57	2.54	5.02	7.51	10.01

Table 9.3   Alpha versus others: macro-system level productivity

Scenario	Min. productivity	Average productivity	Max. productivity	Variance
Alpha	17.569	22.717	24.083	0.6003078
Beta	0.249	0.250	0.250	0.0000000
Gamma	16.195	16.805	17.209	0.0430414
Iota	1.395	1.416	1.449	0.0001804
Theta	20.002	22.829	24.175	0.7206033

forward: i) the Beta scenario tests the generic importance of knowledge in determining the dynamics of productivity and production – we expect the poorest results from the Beta scenario; ii) the Gamma scenario will negate our hypothesis if its results are close to those from the Alpha scenario; and iii) the Iota scenario underlines the dramatic importance of spillovers for the growth of knowledge and productivity. We observe that the three alternative scenarios do not overtake the performance of the Alpha scenario where knowledge externalities are fully at work. Table 9.3 shows the average results of 100 simulations for each scenario: the evidence confirms that the results are not dependent on random distribution due to the meaningless level of variance among the 100 trials even when they were based on different random seeded distributions.

After 2000 production cycles, in the Alpha scenario the system, as a collection of commons, reaches an average productivity of 22.717; in the same number of simulation steps, in the Gamma, Iota and Beta scenarios the system reaches, respectively, 16.805, 1.416 and 0.25. The Theta scenario, based on the same parameter configuration of Alpha, differs only in the number of commons: its results are very close to those of Alpha. The number of commons seems to have very little influence on the results at the macro-system level.

*Table 9.4   Alpha versus others: commons minimum and maximum
            population and turnover*

Scenario	Min. size	Max. size	Min. turnover	Max. turnover
Alpha	24	265	3768	9695
Beta	0	529	145	1190
Gamma	27	277	7054	13,319
Iota	66	141	29,095	31,555
Theta	57	537	11,501	24,440

A batch of 100 simulations was run for the Alpha and Gamma scenarios, with a higher cost of labour (wages were set to 10 instead of 1) in order to test that the Alpha scenario led to higher productivity levels than the Gamma scenario, independent of labour cost. Comparison of these two scenarios confirms the importance of knowledge externalities; even the distance between the final productivity achieved by the two scenarios was less: whereas the Alpha scenario reached an average productivity of 20.585 (over the 100 simulations) the Gamma scenario stopped at 16.626.

Table 9.4 reports the minimum and maximum population achieved during the first 2000 production cycles across the ten commons, as well as the dynamic due to moving across commons by means of minimum and maximum turnover – that is, the sum of enterprises that had entered and exited the commons.

Again, the interpretation is straightforward: the structure of the system is endogenous. There is a clear technological and structural change loop. We see that the pace of productivity at system level is affected by the distribution of firms across commons. At the same time, the structure of the system is affected by the different dynamics of productivity. The loop encompasses historic time and leads to strong non-ergodic path dependence. The Alpha scenario, with strong positive knowledge externalities fully at work, shows lower levels of concentration of firms across commons. Concentrations are greater in the scenarios where the effects of externalities on competence are smaller, and naturally where the number of commons is limited, as in the Theta scenario. Commons-to-commons flows are dramatically higher for the Iota scenario, where firms cannot engage in external learning so react to out-of-equilibrium conditions by moving continuously from commons to commons.

Sensitivity to the key parameters does not raise concern. A few simulations have been devoted to test the sensitivity to four parameters that were expected to have (or might have) a strong influence on the results of the simulations. Table 9.5 briefly summarizes the Pearson's index values com-

Table 9.5   Sensitivity to key parameter values

Scenario	Tolerance	IPR duration	Internal learning	Commons cost
Alpha	0.060	−0.873	0.510	−0.962

puted through 100 simulations run under random values for the following parameters:

- Tolerance – the equilibrium condition; an agent is considered 'in equilibrium' if its results are different from the average more than tolerance, either in negative or positive terms;
- IPR duration – the number of production cycles a technology enhancement is hidden from other firms due to IPR protection;
- Internal learning – the knowledge an enterprise accumulates each step due to learning by doing;
- Commons cost – the base value for commons costs, both knowledge management and exploration.

The Pearson ratios have been computed between the random value of the parameter and the productivity level achieved after 2000 production cycles at the macro-system level, under constant values for each other parameter and the same distribution of the random events.

As Table 9.5 shows, the longer the IPR protection lasts the less productivity the system achieved after 2000 production cycles; the same effect is shown for the Commons costs, but even stronger. A correlation has been found with the learning capability that is a trivial but highly plausible observation. The tolerance level was demonstrated to have a very weak correlation; levels tested were from 0 to 0.001 – the level usually employed for the simulations – in order to demonstrate that even smaller set-ups for this parameter would have added very little to the meaningfulness of the simulations.

The essential observation in Table 9.5 is that the strong correlation between achieved productivity and IPR duration and Commons cost (that is, knowledge governance cost) demonstrates that under poor or null knowledge externalities the behaviour of enterprises is doomed to be simply adaptive. This observation constitutes the ultimate answer to this chapter's research question: the results confirm the claim for the dramatic effects of the endogenous dynamics of knowledge externalities. The analysis of productivity growth in the different scenarios highlights the dramatic gaps between them for average firm output, which is highest in the Alpha scenario.

The Alpha scenario exhibits faster rates of productivity growth and a

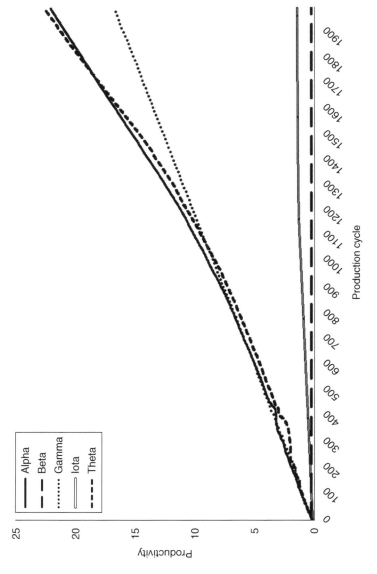

Figure 9.1   Alpha versus others: macro-system level productivity

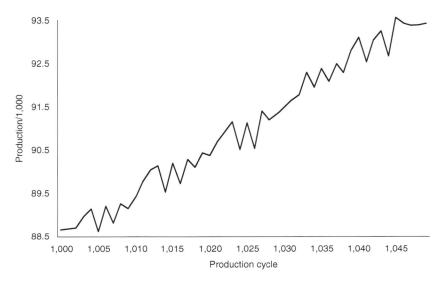

Figure 9.2   Output levels during 50 production cycles

typical step-wise pattern of growth, with periods of fast growth followed by phases of slow growth. Figure 9.1 shows that the availability of net positive knowledge externalities within each commons-cum-mobility of firms across commons, stylized in the Alpha scenario, are able to push the whole economy to far higher productivity values. Consistent with this, output at the macro-system level shows larger growth in the Alpha scenario compared to the others.

Figure 9.2 highlights the typical step-wise pattern of growth when knowledge externalities are fully at work – that is, the Alpha scenario – with periods of fast growth followed by phases of slow growth.

4.2   The Dynamics of the Commons in the Alpha Scenario

As mentioned, the Alpha scenario represents our benchmark, validated by the results of the simulations. It is interesting to explore the dynamics of structural change engendered by the model of creative response-cum-knowledge externalities at the commons level.

At commons level, the results of the simulations show that, although selection of a new commons is a blind activity for agents, their mobility strongly affects the structure of the system and the size of each commons. Figure 9.3 provides a general representation of the phenomenon at commons level by showing the number of firms in the first three commons

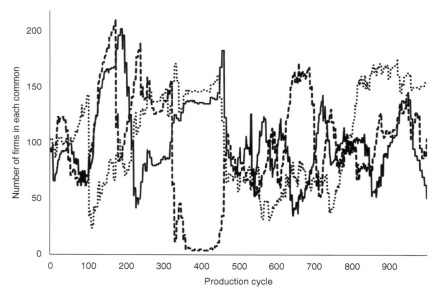

Figure 9.3   Alpha waves of population patterns across commons

during 1000 production cycles. It shows clearly that each commons undergoes a typical Schumpeterian wave, with phases of growth and subsequent decline along the process. The long-term pattern of growth is punctuated by waves where, after rapid take-off, the commons enters a contraction phase due to the rising knowledge governance costs for excessive crowding. As one commons contracts, others increase in size – of output and number of firms.

Over the long term, the oscillations level out and the size of commons becomes increasingly homogeneous, with a clear decline in concentration. Variety among commons seems to exert a strong and positive effect on the overall increase in productivity at system level. This evidence warrants further analysis, but could be considered to hint at the powerful effects of replicator dynamics according to which the rate of growth of a system is positively influenced by its variety (Metcalfe, 2002).

The Schumpeterian waves at commons level affect overall aggregate patterns of productivity growth at system level, which show a typical step-wise pattern (see Figure 9.2). The evidence from these simulations hints at an innovation process conceived as a Schumpeterian creative reaction enabled by knowledge externalities, engendering structural change and 'disorder' at commons level, with marked Schumpeterian waves of output growth and firm populations which positively affect the system-level dynamics where both output and productivity show continuous step-wise growth. Creative destruction occurs at the firm and commons levels, but benefits the system at large. The locus of innovation shifts along time from one commons to another in a punctuated sequence that closely parallels the long-term historic trends identified by Mokyr (1990a).

5.   CONCLUSIONS AND POLICY IMPLICATIONS

The understanding of the pervasive role of the Arrovian properties of knowledge as an economic good – non-appropriability, non-exhaustibility, and cumulability and complementarity stemming from its indivisibility – makes it possible to grasp the recombinant character of its generation process. This process involves external knowledge as an indispensable input in the generation of new knowledge and the eventual introduction of innovations. The creative reaction of firms caught in out-of-equilibrium conditions and the necessary generation of knowledge are enabled by the pecuniary knowledge externalities stemming from the quality and structure of the networks of synchronic and diachronic complementarities among firms linked by formal and informal ties. However, pecuniary knowledge externalities are not always available. The success of the creative reactions of firms caught in out-of-equilibrium conditions to generate new technological knowledge and introduce productivity enhancing innovations depends on the availability of pecuniary knowledge externalities.

In these intrinsically localized circumstances, innovation is a highly specific and idiosyncratic emerging property that takes place only when the complexity of the local system is properly organized and adequate levels of knowledge connectivity are reached and maintained. The success of such creative reactions, in turn, changes the organization of the system and its knowledge connectivity, and may reinforce the availability of pecuniary knowledge externalities, feeding a self-sustained process of growth and change as well as endangering it. The process is far from deterministic: excess density, with the consequent decline of knowledge connectivity, in fact, is a possible outcome of the generation of additional knowledge and the changes in the structure of the system that stem from the introduction of innovations.

Knowledge externalities are endogenous: there is a causal loop linking the amount of knowledge that each firm can generate with the cost of available external knowledge, including knowledge governance costs, which, in turn, depend on the – changing – structure of interactions and transactions, and density of co-localized firms. The larger the pecuniary knowledge externalities, the stronger are the incentives for firms to try to enter knowledge-rich commons. Their entry affects the knowledge connectivity of the system, and hence its knowledge governance costs as well as the supply of technological spillovers, and changes the level of the available pecuniary knowledge externalities.

The stock of external knowledge available at any point in time, and in regional and technological space, is not determined by exogenous factors, but is strongly influenced by the conditions of knowledge governance costs within the knowledge commons, as well as by the amount of creative reactions that have been taking place at each point in time.

The use of an ABM allows us to articulate the relations between the basic ingredients of the dynamic processes, and to elaborate a coherent analytical framework that helps explain, and mimics, the endogenous long-term dynamics of technological and structural change that are at the heart of economic growth. Thus, the ABM can be considered a type of artificial cliometrics, providing the opportunity to test a set of hypotheses about the role of endogenous knowledge externalities. The results of the ABM confirm that endogenous knowledge externalities have powerful effects on the equilibrium conditions of the system dynamics at the micro-, meso- and macro-levels.

At the micro-level we show that the reaction of firms caught in out-of-equilibrium conditions yields successful effects, with the introduction of productivity-enhancing innovations, when pecuniary knowledge externalities provided by high levels of knowledge connectivity are available. Innovation is the result of matching individual and intentional learning efforts in reactive agents with the characteristics of the system in which the firm is embedded. Innovation is an emerging property of the system in which individual action is as indispensable as the availability of positive pecuniary knowledge externalities. Endogenous knowledge externalities generate endogenous growth characterized intrinsically by an out-of-equilibrium state. The introduction of innovation affects the transient equilibrium of product and factor markets, exposes each firm to changes in its relative profitability and induces new innovation efforts. Equilibrium occurs only if and when innovation is impossible because of lack of pecuniary knowledge externalities. Innovation and equilibrium are antithetical.

At the meso-level, the out-of-equilibrium dynamics of endogenous knowledge externalities affect the structural characteristics of the commons and the aggregate system. Endogenous centrifugal and centripetal forces continually reshape each commons and the structure of the system, and produce ever-changing heterogeneity characterized by the creation and decline of knowledge commons. The process exhibits the typical traits of a third-order emergence where microprocesses lead to aggregate changes that, in turn, may affect the likelihood of the microdynamics (Martin and

Sunley, 2012). To try to access pecuniary knowledge externalities, firms can move across commons. This mobility may have the twin effect to: a) increase their chances to innovate; and b) change the structural landscape and the consequent levels of knowledge connectivity of each commons and, hence, of the system, viewed as a collection of commons. Within commons, mobility across commons affects local knowledge governance costs and changes the levels of pecuniary knowledge externalities and, thus, the likelihood that co-localized firms can generate new technological knowledge and introduce technological innovations that will increase their productivity. A knowledge commons endowed with firms that enjoy high levels of productivity may attract many learning firms willing to improve their productivity. Their entry, however, may affect the local levels of knowledge governance costs and reduce the levels of net positive pecuniary knowledge externalities, reducing the overall attractiveness of the location and the aggregate dynamics of the system. Local systems may experience a transition from high levels of organized complexity able to generate high levels of net positive knowledge externalities, to low levels of organized complexity where congestion and governance costs make access to knowledge spillovers more expensive.

At the single commons level, the out-of-equilibrium process leads to non-linear patterns of economic growth characterized by significant oscillations in the firm population levels, and rates of output, profitability and productivity growth that take the form typical of long waves in Schumpeterian analyses of business cycles.

At the system level, the dynamics of productivity growth exhibits a typical step-wise pattern, with long periods of time characterized by smooth rates of increase and sudden, sharp jumps. When the distribution of firms within the knowledge commons is particularly effective and the local system is able to promote high levels of knowledge externalities, the rate of generation of new knowledge and the rate of productivity-enhancing innovation increase. At the aggregate level, the system experiences fast rates of output and productivity growth. In the opposite case, the distribution of firms across knowledge commons reduces the opportunities to benefit from net positive knowledge externalities. Crowded knowledge commons command high levels of knowledge governance costs; peripheral knowledge commons with low levels of productivity involve few opportunities for knowledge dissemination, and the system experiences low rates of innovation introduction and productivity growth.

The endogenous dynamics of knowledge externalities engenders multiple equilibria as well as micro–macro feedback such that the dynamics of the system becomes very sensitive to small and unintended shocks. In the case of a single attractor, prices perform as vectors of reliable signals about market conditions, and competition restores the equilibrium conditions. In the opposite case, in a dynamic context based on out-of-equilibrium conditions, the consequences of individual action on the structural characteristics of the system are difficult to foresee. In the local context, and over a short time span, only procedural rationality will apply. There is no countervailing force that can identify a real attractor. Therefore, entrepreneurial action may have major consequences at the economic system level, with either positive or negative effects. Access to external knowledge and dissemination of knowledge generally are far from automatic. They are stochastic not deterministic processes and may or may not occur, depending on the characteristics of the system that are not given only once and are not exogenous, but rather are constantly changing through time as a consequence of agents' actions.

The endogenous dynamics of pecuniary knowledge externalities is intrinsically path dependent. The existing structure of the system affects the dynamics, but at each point in time firms can change the amount of resources invested in the generation of knowledge; new governance mechanisms can be introduced; and the mobility of firms across the knowledge and regional space changes the structure of the system and the levels of pecuniary knowledge externalities.

The policy implications of these results are important in highlighting the endogenous dynamics of knowledge externalities. Knowledge externalities do not fall like manna from heaven, and are not given once and forever. There has been an extreme focus on knowledge generating policies to the detriment of policies for knowledge governance. Careful policy interventions to promote intentional changes to the system parameters in order to improve knowledge governance could have long-lasting and positive effects (Ostrom and Hess, 2006; Ostrom, 2010).

Knowledge dissemination should become the topic of dedicated policies aimed at favouring the access and use of external knowledge as an indispensable input in the successful recombinant generation of new technological knowledge. The design of specific applications of new IPR regimes that favour knowledge dissemination and yet enable appropriate levels of knowledge appropriability could be very effective for knowledge dissemination. Systematic introduction of measures that would reduce exclusive property rights based on compulsory licensing with fair royalties would likely have strong positive effects on the rates of generation of new technological knowledge. We would stress here that, although the implementation of interventions affecting the basic architecture of IPR regimes might seem rather controversial, our argument becomes more realistic and palatable when considered as the introduction of non-exclusive IPR for patents stemming from public interventions, ranging from public procurement to research activities supported by public funding. Interventions that support the purchase of patents and, more generally, interactions between knowledge producers and knowledge users and all public subsidies would help the dissemination of knowledge both within core regions and among regions (Reichman, 2000).

Support for mobility of skilled personnel can be a very effective tool for knowledge dissemination. In core regions it would help reduce knowledge absorption costs, and across regions support for mobility of skilled personnel, academics and inventors from core regions can make (re)location in a semi-core region more attractive. This type of support would favour interregional knowledge dissemination from core to non-core regions.

The strengthening of effective interactions between firms and the academic system both within and across commons is likely to reduce substantially the costs of external knowledge. The entry of the public research infrastructure in the market for knowledge outsourcing can help the effective absorption of knowledge generated by the public research infrastructure, increasing the actual amount of net positive knowledge externalities. The public research infrastructure can take advantage of the signals provided by firms so as to better direct the internal interdisciplinary allocation of resources. Firms can access the research capabilities of large and effective public R&D labs to perform R&D activities taking advantage of substantial increasing returns and low unit fixed costs. The implementation of effective systems of interaction can improve the matching between the public research infrastructure and the business community so as to increase the amount of net positive knowledge externalities available in the system favouring its growth dynamics.

Support for the creation of academic networks between strong academic institutes in core regions linked by strong institutional ties and peripheral universities located in non-core regions would help the dissemination of knowledge across regions. Within regions, dissemination of academic knowledge would be increased by reducing the exclusivity in academic employment contracts to allow individual academics to participate in knowledge-consulting activities. Especially for small firms, using academics as consultants would allow them to build contractual relations in knowledge typical of large corporations. All interventions that increase access to external knowledge and reduce knowledge interaction costs are likely to exert positive effects on the dynamics of economic systems.

NOTES

1. We define the equilibrium cost of knowledge as the cost of a standard divisible input, traded in a competitive market, that can be fully appropriated, that is subject to wear and tear because of high levels of exhaustibility and obsolescence, and that is used to produce an output that is traded in a competitive market.
2. NK models assume the reverse, defining density of the components in the landscape and their knowledge to be exogenous (Levinthal, 1997).
3. See Antonelli and Ferraris (2011, 2017a) for complementary specifications of this model.
4. See Antonelli and Ferraris (2011, 2017a) for other complementary specifications of this simulation model.
5. Note that the system is analytically consistent. Naming Π the profit of a generic enterprise and D the dividend it will pay to its shareholders, and remembering equations (9.1), (9.2) and (9.5), it is possible to write the following equations:

$$D_i = \Pi_i = pO_i - W_i \qquad (9.15)$$

where D could be less than zero if a loss had to be reintegrated. The amount of dividends paid to the whole system is:

$$D = \sum D_i \qquad (9.16)$$

At the aggregate level the system could be summarized as follows:

$$Y = \sum W_i + \sum D_i \qquad (9.17)$$

By specifying D_i using equation (9.16) it is possible to obtain:

$$Y = \sum W_i + \sum pO_i - \sum W_i \qquad (9.18)$$

By operating simple compensations, equation (9.18) becomes:

$$Y = \sum pO_i \qquad (9.19)$$

Recalling equation (9.2), it is evident that the whole system can reach equilibrium and the amount of money in the system remains constant.

References

Abramovitz, M. (1956), 'Resources and output trends in the US since 1870', *American Economic Review*, **46**, 5–23.

Abramovitz, M. (1989), *Thinking about Growth*, Cambridge: Cambridge University Press.

Abramovsky, L. and Griffith, R. (2006), 'Outsourcing and offshoring of business services: How important is ICT?', *Journal of the European Economic Association*, **4** (2–3), 594–601.

Acemoglu, D., Aghion, P., Bursztyn, L. and Hémous, D. (2012), 'The environment and directed technical change', *American Economic Review*, **102**, 131–66.

Adams, J.D. (1990), 'Fundamental stocks of knowledge and productivity growth', *Journal of Political Economy*, **98**, 673–702.

Adams, J. (2006), 'Learning internal research and spillovers', *Economics of Innovation and New Technology*, **15**, 5–36.

Aghion, P., Dechezlepretre, A., Hémous, D., Martin, R. and Van Reenen, J. (2016), 'Carbon taxes, path dependency, and directed technical change: Evidence from the auto industry', *Journal of Political Economy*, **124**, 52–104.

Aghion, P., David, P.A. and Foray, D. (2009), 'Science, technology and innovation for economic growth: Linking policy research and practice in "STIG Systems"', *Research Policy*, **38**, 681–93.

Aghion, P. and Howitt, P. (1992), 'A model of growth through creative destruction', *Econometrica*, **60**, 323–51.

Aghion, P. and Howitt, P. (1998), *Endogenous Growth Theory*, Cambridge, MA: MIT Press.

Ahn, S. (2000), 'Firm dynamics and productivity growth: A review of micro evidence from OECD countries', OECD Economic Department WP 297.

Alchian, A. (1950), 'Uncertainty evolution and economic theory', *Journal of Political Economy*, **58**, 211–21.

Amiti, M. and Wei, S. (2009), 'Service offshoring and productivity: Evidence from the US', *World Economy*, **32** (2), 203–20.

Anderson, P.W., Arrow, K.J. and Pines, D. (eds) (1988), *The Economy as an Evolving Complex System*, Reading, MA: Addison-Wesley.

Angrist, J.D. and Pischke, J.S. (2008), *Mostly Harmless Econometrics*, Princeton, NJ: Princeton University Press.

Antonelli, C. (1997), 'The economics of path-dependence in industrial organization', *International Journal of Industrial Organization*, **15**, 643–75.

Antonelli, C. (1999), *The Microdynamics of Technological Change*, London: Routledge.

Antonelli, C. (2008), *Localized Technological Change: Towards the Economics of Complexity*, London: Routledge.

Antonelli, C. (2009), 'The economics of innovation: From the classical legacies to the economics of complexity', *Economics of Innovation and New Technology*, **18**, 611–46.

Antonelli, C. (ed.) (2011), *Handbook on the Economic Complexity of Technological Change*, Cheltenham, UK and Northampton, MA, USA: Edward Elgar Publishing, pp. 1–62.

Antonelli, C. (2013a), 'Knowledge governance, pecuniary knowledge externalities and total factor productivity growth', *Economic Development Quarterly*, **27**, 62–70.

Antonelli, C. (2013b), 'Compulsory licensing: The foundations of an institutional innovation', *WIPO Journal*, **4**, 157–74.

Antonelli, C. (2015a), 'Innovation as a creative response: A reappraisal of the Schumpeterian legacy', *History of Economic Ideas*, **23**, 99–118.

Antonelli, C. (2015b), 'The dynamics of knowledge governance', in C. Antonelli and A. Link (eds), *Handbook on the Economics of Knowledge*, London: Routledge, pp. 232–62.

Antonelli, C. (2017a), 'Endogenous innovation: The creative response', *Economics of Innovation and New Technology*, **26** (8), 689–718.

Antonelli, C. (2017b), *Endogenous Innovation: The Economics of an Emergent System Property*, Cheltenham, UK and Northampton, MA, USA: Edward Elgar Publishing.

Antonelli, C. and Colombelli, A. (2015a), 'External and internal knowledge in the knowledge generation function', *Industry and Innovation*, **22**, 273–98.

Antonelli, C. and Colombelli, A. (2015b), 'The knowledge cost function', *International Journal of Production Economics*, **168**, 290–302.

Antonelli, C. and David, P.A. (eds) (2016), *The Economics of Knowledge and Knowledge Driven Economy*, London: Routledge.

Antonelli, C. and Fassio, C. (2016), 'The role of external knowledge(s) in the introduction of product and process innovations', *R&D Management*, **46**, 979–91.

Antonelli, C. and Ferraris, G. (2011), 'Innovation as an emerging system property: An agent based model', *Journal of Artificial Societies and Social Simulation*, **14** (2), DOI: 10.18564/jasss.1741.

Antonelli, C. and Ferraris, G. (2017a), 'The Marshallian and Schumpeterian microfoundations of evolutionary complexity: An agent based simulation model', in A. Pyka and U. Cantner (eds), *Foundations of Economic Change: A Schumpeterian View on Behaviour, Interaction and Aggregate Outcomes*, Heidelberg, Berlin and New York: Springer, pp. 461–500.

Antonelli, C. and Ferraris, G. (2017b), 'The creative response and the endogenous dynamics of pecuniary knowledge externalities: An agent based simulation model', *Journal of Economic Interaction and Coordination*, https://doi.org/10.1007/s11403-017-0194-3.

Antonelli, C. and Gehringer, A. (2016), 'The cost of knowledge and productivity dynamics: An empirical investigation on a panel of OECD countries', in A.N. Link and C. Antonelli (eds), *Strategic Alliances: Leveraging Economic Growth and Development*, London: Routledge, pp. 155–74.

Antonelli, C. and Scellato, G. (2011), 'Out-of-equilibrium profit and innovation', *Economics of Innovation and New Technology*, **20** (5), 405–21.

Antonelli, C. and Scellato, G. (2013), 'Complexity and innovation: Social interactions and firm level productivity growth', *Journal of Evolutionary Economics*, **23**, 77–96.

Antonelli, C. and Scellato, G. (2015), 'Firms size and directed technical change', *Small Business Economics*, **44** (1), 207–18.

Antonelli, C., Barbiellini Amidei, F. and Fassio, C. (2014), 'The mechanisms of knowledge governance: State owned corporations and Italian economic growth, 1950–1994', *Structural Change and Economic Dynamics*, **31**, 43–63.

Antonelli, C., Crespi, F. and Scellato, G. (2012), 'Inside innovation persistence: New evidence from Italian micro-data', *Structural Change and Economic Dynamics*, **23** (4), 341–53.

Antonelli, C., Crespi, F. and Scellato, G. (2013), 'Internal and external factors in innovation persistence', *Economics of Innovation and New Technology*, **22**, 256–80.

Antonelli, C., Crespi, F. and Scellato, G. (2015), 'Productivity growth persistence: Firm strategies, size and system properties', *Small Business Economics*, **45** (1), 129–47.

Antonelli. C., Krafft, J. and Quatraro, F. (2010), 'Recombinant knowledge and growth: The case of ICTs', *Structural Change and Economic Dynamics*, **21**, 50–69.

Antonelli, C, Patrucco, P.P. and Quatraro, F. (2011), 'Productivity growth and pecuniary knowledge externalities: An empirical analysis of agglomeration economies in European regions', *Economic Geography*, **87**, 23–50.

Aoki, M. and Yoshikawa, H. (2002), 'Demand saturation: Creation and economic growth', *Journal of Economic Behavior & Organization*, **48** (2), 127–54.

Arora, A. and Gambardella, A. (1990), 'Complementarity and external linkages: The strategies of the large firms in biotechnology', *Journal of Industrial Economics*, 38, 361–79.

Arrighetti, A., Landini, F. and Lasagni, A. (2014), 'Intangible assets and firm heterogeneity: Evidence from Italy', *Research Policy*, **43**, 202–13.

Arrow, K.J. (1962), 'Economic welfare and the allocation of resources for invention', in R.R. Nelson [NBER] (ed.), *The Rate and Direction of Inventive Activity: Economic and Social Factors*, Princeton, NJ: Princeton University Press, pp. 609–25.

Arrow, K.J. (1969), 'Classificatory notes on the production and transmission of technical knowledge', *American Economic Review*, **59**, 29–35.

Arthur, W.B. (1989), 'Positive feedbacks in the economy', *Scientific American*, **262**, 92–99.

Arthur, W.B. (2007), 'Complexity and the economy', in H. Hanusch and A. Pyka (eds), *Elgar Companion to Neo-Schumpeterian Economics*, Cheltenham, UK and Northampton, MA, USA: Edward Elgar Publishing, pp. 1102–10.

Arthur, W.B. (2009), *The Nature of Technology: What It Is and How It Evolves*, New York: Free Press.

Arthur, W.B. (2014), *Complexity and the Economy*, New York: Oxford University Press.

Arthur, W.B., Durlauf, S.N. and Lane, D.A. (eds) (1997), *The Economy As an Evolving Complex System II*, Reading, MA: Addison-Wesley.

Atkinson, A.B. and Stiglitz, J.E. (1969), 'A new view of technological change', *Economic Journal*, **79**, 573–78.

Axtell, R. (2005), 'The complexity of exchange', *Economic Journal*, **115**, F193–210.

Baily, M.N., Hulten, C. and Campbell, D. (1992), 'Productivity dynamics in manufacturing plants', Brooking Papers on Economic Activity: Microeconomics, 187–249.

Barabasi, L.A. (2010), *Bursts: The Hidden Pattern behind Everything We Do*, New York: Dutton.

Bartelsman, E. and Dhrymes, P. (1998), 'Productivity dynamics: U.S. manufacturing plants 1972–1986', *Journal of Productivity Analysis*, **9** (1), 5–34.

Bartelsman, E. and Doms, M. (2000), 'Understanding productivity: Lessons from longitudinal microdata', *Journal of Economic Literature*, **38** (3), 569–94.

Bartelsman, E., Haltiwanger, J. and Scarpetta, S. (2009), 'Cross country differences in productivity: The role of allocation and selection', NBER Working Paper 15490.

Bartelsman, E.J., Caballero, R.I. and Lyons, R.K (1994), 'Customer-driven and supplier-driven externalities', *American Economic Review*, **84**, 1075–84.

Beaudry, C. and Breschi, S. (2003), 'Are firms in clusters really more innovative?', *Economics of Innovation and New Technology*, **12**, 325–42.

Belenzon, S. (2012), 'Cumulative innovation and market value: Evidence from patent citations', *Economic Journal*, **122**, 265–85.

Ben Hassine, H., Boudier, F. and Mathieu, C. (2017), 'The two ways of FDI R&D spillovers: Evidence from the French manufacturing industry', *Applied Economics*, **49**, 2395–408.

Bergek, A., Berggren, C., Magnusson, T. and Hobday, M. (2013), 'Technological discontinuities and the challenge for incumbent firms: Destruction, disruption or creative accumulation?', *Research Policy*, **42**, 1210–24.

Bianchi, M., Campo dall'Orto, S., Frattini, F. and Vercesi, P. (2010), 'Enabling open innovation in small and medium-sized enterprises: How to find alternative applications for your technologies', *R&D Management*, **40**, 414–31.

Biatour, B. and Dumont, M. (2011), 'The determinants of industry-level total factor productivity in Belgium', Federal Planning Bureau Working Paper, 7–11.

Bischi, G.I., Dawid, H. and Kopel, M. (2003), 'Gaining the competitive edge using internal and external spillovers: A dynamic analysis', *Journal of Economic Dynamics and Control*, **27**, 2171–93.

Bloom, N. and Van Reenen, J. (2007), 'Measuring and explaining management practices across firms and countries', *Quarterly Journal of Economics*, **122** (4), 1351–408.

Bloom, N. and Van Reenen, J. (2010), 'Why do management practices differ across firms and countries?', *Journal of Economic Perspectives*, **24** (1), 203–24.

Blume, L.E. and Durlauf, S.N. (2005), *The Economy As an Evolving Complex System III*, Oxford: Oxford University Press.

Blume, L.E. and Durlauf, S.N. (eds) (2001), *Social Dynamics*, Cambridge, MA: MIT Press.

Bontempi, M. and Mairesse, J. (2008), 'Intangible capital and productivity: An exploration on a panel of Italian manufacturing firms', NBER Working Papers 14108.

Boschma, R.A. (2005), 'Proximity and innovation: A critical assessment', *Regional Studies*, **39**, 61–74.

Bottazzi G., Secchi A. and Tamagni, F. (2008), 'Productivity, profitability and financial performance', *Industrial and Corporate Change*, **17**, 711–51.

Bou, J.C. and Satorra, A. (2007), 'The persistence of abnormal returns at industry and firm levels: Evidence from Spain', *Strategic Management Journal*, **28**, 707–22.

Breschi, S., Lissoni, F. and Malerba, F. (2003), 'Knowledge relatedness in firm technological diversification', *Research Policy*, **32**, 69–97.

Bresnahan, T., Gambardella, A. and Saxenian, A. (2001), 'Old economy inputs for new economy outputs: Cluster formation in the new Silicon Valleys', *Industrial Corporate Change*, **10**, 835–60.

Broedner, P., Kinkel, S. and Lay, G. (2009), 'Productivity effects of outsourcing: New evidence on the strategic importance of vertical integration decisions', *International Journal of Operations and Production Management*, **29** (2), 127–50.

Brynjolfsson, E. and Hitt, L.M. (2000), 'Beyond computation: Information technology, organizational transformation and business performance', *Journal of Economic Perspectives*, **14** (4), 23–48.

Caldari, K. (2015), 'Marshall and complexity: A necessary balance between process and order', *Cambridge Journal of Economics*, **39**, 1071–85.

Camagni, R. and Capello, R. (2013), 'Regional innovation patterns and the EU regional policy reform: Toward smart innovation policies', *Growth and Change*, **44** (2), 355–89.

Cameron, G., Proudman, J., and Redding, S. (2005), 'Technological convergence, R&D, trade and productivity growth', *European Economic Review*, **49**, 775–807.

Cappelli, R., Czarnitzki, D. and Kornelius, K. (2014), 'Sources of spillovers for imitation and innovation', *Research Policy*, **43** (1), 115–20.

Cassiman, B. and Veugelers, R. (2006), 'In search of complementarity in innovation strategy: Internal R&D and external knowledge acquisition', *Management Science*, **52**, 68–82.

Cefis, E. (2003), 'Is there persistence in innovative activities?', *International Journal of Industrial Organization*, **21**, 489–515.

Cefis, E. and Orsenigo, L. (2001), 'The persistence of innovative activities: A cross-countries and cross-sectors comparative analysis', *Research Policy*, **30**, 1139–58.

Chandler, A.D. (1977), *The Visible Hand: The Managerial Revolution in American Business*, Cambridge, MA: Belknap Press of Harvard University Press.

Chandler, A.D. (1990), *Scale and Scope: The Dynamics of Industrial Capitalism*, Cambridge, MA: Belknap Press of Harvard University Press.

Chesbrough, H. (2003), *Open Innovation: The New Imperative for Creating and Profiting from Technology*, Boston, MA: Harvard Business School Press.

Chesbrough, H. and Crowther, A.K. (2006), 'Beyond high tech: Early adopters of open innovation in other industries', *R&D Management*, **36** (3), 229–36.

Chesbrough, H., Vanhaverbeke, W. and West, J. (2006), *Open Innovation: Researching a New Paradigm*, Oxford: Oxford University Press.

Cincera, M., De Clerq, P. and Maghe, V. (2014), 'First typology of the national innovation systems in the 28 EU member states and in the 9 third countries covered by the ENIRI study', ENIRI Project, Brussels.

Clausen, T., Pohjola, M., Sapprasert, K. and Verspagen B. (2012), 'Innovation strategies as a source of persistent innovation', *Industrial and Corporate Change*, **21**, 553–85.

Coad, A. and Rao, R. (2006), 'Innovation and market value: A quantile regression analysis', *Economics Bulletin*, **15**, 1–10.

Cohen, W.M. (2010), 'Fifty years of empirical studies of innovative activity and performance', in B.H. Hall and N. Rosenberg (eds), *Handbook of the Economics of Innovation*, Amsterdam: Elsevier, pp. 131–213.

Cohen, W.M. and Klepper, S. (1996), 'A reprise of size and R&D', *Economic Journal*, **106** (437), 25–51.

Cohen, W.M. and Levinthal, D.A. (1989), 'Innovation and learning: The two faces of R&D', *Economic Journal*, **99**, 569–96.

Cohen, W.M. and Levinthal, D.A. (1990), 'Absorptive capacity: A new perspective on learning and innovation', *Administrative Science Quarterly*, **35**, 128–52.

Colombelli, A. and von Tunzelmann, G.N. (2011), 'The persistence of innovation and path dependence', in C. Antonelli (ed.), *Handbook on the Economic Complexity of Technological Change*, Cheltenham, UK and Northampton, MA, USA: Edward Elgar Publishing, pp. 105–20.

Colombelli, A., Foddi, M. and Paci, R. (2013a), 'Scientific regions', in R. Capello and C. Lenzi (eds), *Territorial Patterns of Innovation: An Inquiry on the Knowledge Economy in European Regions*, London: Routledge, pp. 43–69.

Colombelli, A., Krafft, J. and Quatraro, F. (2013b), 'Properties of knowledge base and firm survival: Evidence from a sample of French manufacturing firms', *Technological Forecasting and Social Change*, **80**, 1469–83.

Colombelli, A., Krafft, J. and Quatraro, F. (2014), 'High growth firms and technological knowledge: Do gazelles follow exploration or exploitation strategies?', *Industrial and Corporate Change*, **23** (1), 261–91.

Conte, A. and Vivarelli, M. (2005), 'One or many knowledge production functions? Mapping innovative activity using microdata', IZA DP No. 1878.

Corrado, C., Hulten, C. and Sichel, D. (2009), 'Intangible capital and economic growth', *Review of Income and Wealth*, **55** (3), 661–85.

Cowan, R. and Jonard, N. (2004), 'Network structure and the diffusion of knowledge', *Journal of Economic Dynamics and Control*, **28**, 1557–75.

Cowan, R., Jonard, N. and Zimmermann, J.B. (2006), 'Evolving networks of inventors', *Journal of Evolutionary Economics*, **16**, 155–74.

Cowan, R., Jonard, N. and Zimmermann, J.B. (2007), 'Bilateral collaboration and the emergence of networks', *Management Science*, **53**, 1051–67.

Crépon, B., Duguet, E. and Mairesse, J. (1998), 'Research and development, innovation and productivity: An econometric analysis at the firm level', *Economics of Innovation and New Technology*, **7** (2), 115–58.

Crespi, F. (2007), 'IT services and productivity in European industries', in H.L.M. Kox and L. Rubalcaba (eds), *Business Services in European Economic Growth*, London: Palgrave Macmillan.

Crespi, F. and Scellato, G. (2014), 'Knowledge cumulability and path dependence in innovation persistence', in C. Antonelli and A. Link (eds), *The Routledge Handbook of the Economics of Knowledge*, London: Routledge, pp. 116–34.

Cyert, R.M. and March, J.C. (1963), *A Behavioral Theory of the Firm*, Englewood Cliffs, NJ: Prentice-Hall.

D'Ignazio, A. and Giovannetti, E (2006), 'From exogenous to endogenous economic networks: Internet applications', *Journal of Economic Survey*, **20**, 757–96.

Da Silva, M.A. (2014), 'The knowledge multiplier', *Economics of Innovation and New Technology*, **23** (7), 652–88.

Dahlander, L. and Gann, D.M. (2010), 'How open is innovation?', *Research Policy*, **39**, 699–709.

Dasgupta, P. and Stiglitz, J. (1980), 'Industrial structure and the nature of innovative activity', *Economic Journal*, **90**, 266–93.

David, P.A. (2007), 'Path dependence: A foundational concept for historical social science', *Cliometrica: Journal of Historical Economics and Econometric History*, **1**, 91–114.

David, P.A. (1985), 'Clio and the economics of QWERTY', *American Economic Review*, **75**, 332–37.

David, P.A. (1993), 'Knowledge property and the system dynamics of technological change', *Proceedings of the World Bank Annual Conference on Development Economics*, Washington, DC: World Bank.

David, P.A. and Keely, L. (2003), 'The endogenous formation of scientific research coalitions', *Economics of Innovation and New Technology*, **12**, 93–116.

Dawid, H. (2006), 'Agent-based models of innovation and technical change', in L. Tesfatsion and K.L. Judd (eds), *Handbook of Computational Economics, 2: Agent-Based Computational Economics*, Amsterdam: North-Holland, pp. 1235–72.

Deschryvere M. (2014), 'R&D, firm growth and the role of innovation persistence: An analysis of Finnish SMEs and large firms', *Small Business Economics*, **43** (4), 767–85.

Dobusch, L. and Schüler, E. (2013), 'Theorizing path dependence: A review of positive feedback mechanisms in technology markets, regional clusters, and organizations', *Industrial and Corporate Change*, **22**, 617–47.

Di Stefano, G., Gambardella, A. and Verona, G. (2012), 'Technology push and demand pull perspectives in innovation studies: Current findings and future research directions', *Research Policy*, **41**, 1283–95.

Dosi, G. (1982), 'Technological paradigms and technological trajectories: A suggested interpretation of the determinants and directions of technical change', *Research Policy*, **11**, 147–62.

Dosi, G. (1997), 'Opportunities, incentives and the collective patterns of technological change', *Economic Journal*, **107** (444), 1530–47.

Dosi, G., Marsili, O., Orsenigo, L. and Salvatore, R. (1995), 'Learning, market selection and the evolution of industrial structures', *Small Business Economics*, **7** (6), 411–36.

Durlauf, S.N. (2005), 'Complexity and empirical economics', *Economic Journal*, **115**, 225–43.

Ebersberger, B. and Herstad, S.J. (2011), 'Product innovation and the complementarities of external interfaces', *European Management Review*, **8**, 117–35.

Engelsman, E.C. and van Raan, A.F.J. (1994), 'A patent-based cartography of technology', *Research Policy*, **23**, 1–26.

Enkel, E., Gassman, O. and Chesbrough, H. (2009), 'Open R&D and open innovation: Exploring the phenomenon', *R&D Management*, **39**, 331–41.

Erixon L. (2016), 'Is firm renewal stimulated by negative shocks? The status of negative driving forces in Schumpeterian and Darwinian economics', *Cambridge Journal of Economics*, **40**, 93–121.

Faems, D., De Visser, M., Andries, P. and Van Looy, B. (2010), 'Technology alliance portfolios and financial performance: Value-enhancing and cost-increasing effects of open innovation', *Journal of Product Innovation Management*, **27**, 785–96.

Faggio, G., Salvanes, K.G. and Van Reenen, J. (2009), 'The evolution of inequality in productivity and wages: Panel data evidence', LSE Working Paper.

Fassio, C. (2015), 'How similar is innovation in German, Italian, and Spanish medium-technology sectors? Implications for the sectoral systems of innovation and distance-to-the-frontier perspectives', *Industry and Innovation*, **22** (2) 102–25.

Feldman, M.A. (1999), 'The new economics of innovation, spillovers and agglomeration: A review of empirical studies', *Economics of Innovation and New Technology*, **8**, 5–25.

Feldman, M.A. (2003), 'The locational dynamics of the US biotech industry: Knowledge externalities and the anchor hypothesis', *Industry and Innovation*, **10** (3), 311–29.

Fleming, L. (2001), 'Recombinant uncertainty in technological search', *Management Science*, **47** (1), 117–32.

Fleming, L. and Sorenson, O. (2001), 'Technology as a complex adaptive system: Evidence from patent data', *Research Policy*, **30**, 1019–39.

Foster, J. (2005), 'From simplistic to complex systems in economics', *Cambridge Journal of Economics*, **29**, 873–92.

Foster, J. and Metcalfe, J.S. (2012), 'Economic emergence: An evolutionary economic perspective', *Journal of Economic Behavior and Organization*, **82** (2), 420–32.

Foster, L., Haltiwanger, J. and Krizan, C.J. (2001), 'Aggregate productivity growth: Lessons from microeconomic evidence', in E. Dean, M. Harper and C. Hulten (eds), *New Developments in Productivity Analysis*, Chicago: University of Chicago Press, pp. 303–72.

Fritsch, M. (2002), 'Measuring the quality of regional innovation systems: A knowledge production function approach', *International Regional Science Review*, **25**, 86–101.

Furman, J.L., Porter, M.E. and Stern, S. (2002), 'The determinants of national innovative capacity,' *Research Policy*, **31**, 899–933.

Galindo-Rueda, F. and Haskel, J. (2005), 'Skills, workforce characteristics and firm-level productivity in England', Report prepared for the Department of Trade and Industry, Department for Education and Skills, Office for National Statistics.

Gay, C., Latham, W. and Le Bas, C. (2008), 'Collective knowledge, prolific inventors and the value of inventions: An empirical study of French, German and British patents in the US, 1975–1999', *Economics of Innovation and New Technology*, **17**, 5–22.

Gehringer A. (2011a), 'Pecuniary knowledge externalities across European countries: Are there leading sectors?', *Industry and Innovation*, **18** (4), 415–36.

Gehringer, A. (2011b), 'Pecuniary knowledge externalities and innovation: Intersectoral linkages and their effects beyond technological spillovers', *Economics Innovation and New Technology*, **20**, 495–515.

Gehringer, A. (2012), 'A new sectoral taxonomy based on pecuniary knowledge externalities: Knowledge interactions in a vertically integrated system', *Economic System Research*, **24**, 35–55.

Gehringer, A. (2013), 'Financial liberalization, growth, productivity and capital accumulation: The case of European integration', *International Review of Economics and Finance*, **25**, 291–309.

Gehringer, A., Martínez-Zarzoso, I. and Nowak-Lehmann Danzinger, F. (2016a), 'TFP estimation and productivity drivers in the European Union', CEGE Discussion Paper 189.

Gehringer, A., Martínez-Zarzoso, I. and Nowak-Lehmann Danzinger, F. (2016b), 'What are the drivers of total factor productivity in the European Union?', *Economics of Innovation and New Technology*, **25** (4) 406–34.

Geroski, P., Kretschmer, T. and Walters, C. (2009), 'Corporate productivity growth: Leaders and laggards', *Economic Enquiry*, **47** (1), 1–17.

Geroski, P., Lazarova, S., Urga, G. and Walters, C.F. (2003), 'Are difference in firm size transitory or permanent?', *Journal of Applied Econometrics*, **18**, 47–59.

Giannangeli, S. and Gomez-Salvador, R. (2008), 'Evolution and sources of manufacturing productivity growth: Evidence from a panel of European Countries', European Central Bank: Working Paper Series.

Gilley, K. and Rasheed, A. (2000), 'Making more by doing less: An analysis of outsourcing and its effects on firm performance', *Journal of Management*, **26** (4), 763–90.

Gomulka, S. (1970), 'Extensions of "the golden rule of research" of Phelps', *Review of Economic Studies*, **37**, 73–93.

Griffith, R., Harrison, R. and Van Reenen, J. (2006a), 'How special is the special relationship? Using the impact of U.S. R&D spillovers on U.K. firms as a test of technology sourcing', *American Economic Review*, **96**, 1859–75.

Griffith, R., Huergo, E., Mairesse, J. and Peters, B. (2006b), 'Innovation and productivity across four European countries', *Oxford Review of Economic Policy*, **22** (4), 483–98.

Griffith, R., Redding, S. and Van Reenan, J. (2003), 'R&D and absorptive capacity: Theory and empirical evidence', *Scandinavian Journal of Economics*, **105** (1), 99–118.

Griliches, Z. (1979), 'Issues in assessing the contribution of research and development to productivity growth', *Bell Journal of Economics*, **10** (1), 92–116.

Griliches, Z. (1990), 'Patent statistics as economic indicators: A survey', *Journal of Economic Literature*, **28**, 1661–707.

Griliches, Z. (1992), 'The search for R&D spillovers', *Scandinavian Journal of Economics*, **94** (Supplement), 29–47.

Griliches, Z. (ed.) (1984), *R&D Patents and Productivity*, Chicago: University of Chicago Press.

Grillitsch, M., Tödtling, F. and Höglinger, C. (2013), 'Variety in knowledge sourcing, geography and innovation: Evidence from the ICT sector in Austria', *Papers in Regional Science*, **94**, 25–43.

Grimpe, C. and Kaiser, U. (2010), 'Balancing internal and external knowledge acquisition: The gains and pains from R&D outsourcing', *Journal of Management Studies*, **47** (8), 1483–509.

Grossman, G. and Helpman, E. (1991), *Innovation and Growth in the Global Economy*, Cambridge, MA: MIT Press.

Grossman, G. and Helpman, E. (2005), 'Outsourcing in a global economy', *Review of Economic Studies*, **72** (250), 135–59.

Guiso, L. and Schivardi, F. (2007), 'Spillovers in industrial districts', *Economic Journal*, **117** (516), 68–93.

Gunday, G., Ulusoy, G., Kilic, K. and Alpkan, L. (2011), 'Effects of innovation types on firm performance', *International Journal of Production Economics*, **133** (2), 662–76.

Hall, B.H. and Mairesse, J. (1995), 'Exploring the relationship between R&D and productivity in French manufacturing firms', *Journal of Econometrics*, **65**, 263–93.

Hall, B.H., Jaffe, A.B. and Trajtenberg, M. (2005), 'Market value and patent citations', *Rand Journal of Economics*, **36**, 16–38.

Harison, E. (2008), 'Intellectual property rights in a knowledge-based economy: A new frame-of-analysis', *Economics of Innovation and New Technology*, **17**, 377–400.

Harper, D.A. and Lewis, P. (2012), 'New perspectives on emergence in economics', *Journal of Economic Behavior & Organization*, **82**, 329–37.

Heckman, J.J. (1981), 'The incidental parameters problem and the problem of initial conditions in estimating a discrete time-discrete data stochastic process', in C.F. Manski and D. McFadden (eds), *Structural Analysis of Discrete Data with Econometric Applications*, Cambridge, MA: MIT Press, pp. 179–95.

Henrekson, M. and Johansson, D. (2010), 'Gazelles as job creators: A survey and interpretation of the evidence', *Small Business Economics*, **35**, 227–44.

Heshmati, A. (2003), 'Productivity growth, efficiency and outsourcing in manufacturing and service industries', *Journal of Economic Surveys*, **17** (1), 79–112.

Hölzl, W. (2009), 'Is the R&D behaviour of fast-growing SMEs different? Evidence from CIS III data for 16 countries', *Small Business Economics*, **33** (1), 59–75.

Howells, J., Gagliardi, D. and Malik, K. (2008), 'The growth and management of R&D outsourcing: Evidence from UK pharmaceuticals', *R&D Management*, **38**, 205–19.

Ilmakunnas, P., Maliranta, M., and Vainiomäki, J. (2004), 'The roles of employer and employee characteristics for plant productivity', *Journal of Productivity Analysis*, **21** (3), 249–76.

Ito, K. and Lechevalier, S. (2010), 'Why some firms persistently outperform others: Investigating the interactions between innovation and exporting strategies', *Industrial and Corporate Change*, 19, 1997–2039.

Iwai, K. (1984), 'Schumpeterian dynamics: An evolutionary model of innovation and imitation', *Journal of Economic Behavior & Organization*, **5** (2), 159–90.

Iwai, K. (2000), 'A contribution to the evolutionary theory of innovation, imitation and growth', *Journal of Economic Behavior & Organization*, **43**, 167–98.

Goya, E., Vayá, E. and Suriñach, J. (2013), 'Do spillovers matter? CDM model estimates for Spain using panel data', SEARCH WP 4/28.

Jaffe, A. (1986), 'Technological opportunity and spillovers of R&D: Evidence from firms' patents, profits, and market value', *American Economic Review*, **76** (5), 984–1001.

Jaffe, A. (1989), 'Real effects of academic research', *American Economic Review*, **79** (5), 957–70.

Johansson, B. and Lööf, H. (2008), 'Innovation activities explained by firm attributes and location', *Economics of Innovation and New Technology*, **16**, 533–52.

Johansson, B., Lööf, H. (2014), 'R&D strategy, metropolitan externalities and productivity: Evidence from Sweden', *Industry and Innovation*, **21**, 141–54.

Johansson, B., Lööf, H. (2015), 'Innovation strategies combining internal and external knowledge', in C. Antonelli and A. Link (eds), *Handbook of the Economics of Knowledge*, London: Routledge, pp. 29–52.

Jones, C.I. (1995), 'R&D based models of economic growth', *Journal of Political Economy*, **103**, 759–84.

Kim, J. and Chang-Yang, L. (2011), 'Technological regimes and the persistence of first-mover advantages', *Industrial and Corporate Change*, **20**, 1305–33.

Kirman, A. (1997), 'The economy as an evolving network', *Journal of Evolutionary Economics*, **7**, 339–53.

Kirman, A. (2011), 'Learning in agent-based models', *Eastern Economic Journal*, **37** (1), 20–27.

Kirman, A. (2016), 'Complexity and economic policy: A paradigm shift or a change in perspective? A review essay on David Colander and Roland Kupers's complexity and the art of public policy', *Journal of Economic Literature*, **54** (2), 534–72.

Knudsen, M.P. and Mortensen, T.B. (2011), 'Some immediate – but negative – effects of openness on product development performance', *Technovation*, **31**, 54–64.

Krafft, J., Quatraro, F. and Saviotti, P.P. (2009), 'The evolution of the knowledge base in biotechnology: Social network analysis of biotechnology', *Economics of Innovation and New Technology*, **20**, 445–75.

Krafft, J., Quatraro, F. and Saviotti, P.P. (2014), 'Knowledge characteristics and the dynamics of technological alliances in pharmaceuticals: Empirical evidence from Europe, US and Japan', *Journal of Evolutionary Economics*, **24** (3), 587–622.

Krugman, P. (1994), 'Complex landscapes in economic geography', *American Economic Review*, **84**, 412–17.

Krugman, P. (1995), *Development Geography and Economic Theory*, Cambridge, MA: MIT Press.

Kuznets, S. (1971), *Economic Growth of Nations: Total Output and Production Structure*, Cambridge, MA: Harvard University Press.

Laitner, J. and Stolyarov, D. (2013), 'Derivative ideas and the value of intangible assets', *International Economic Review*, **54** (1) 59–95.

Lane, D.A. (2002), 'Complexity and local interactions: Towards a theory of industrial districts', in A. Quadrio Curzio and M. Fortis (eds), *Complexity and Industrial Clusters: Dynamics and Models in Theory and Practice*, Heidelberg and New York: Physica-Verlag, pp. 65–82.

Lane, D.A. and Maxfield, R. (1997), 'Foresight complexity and strategy', in W.B Arthur, S.N. Durlauf and D.A. Lane (eds), *The Economy As an Evolving Complex System II*, Reading, MA: Addison-Wesley, pp. 169–98.

Lane, D.A., Pumain, D., Leeuw, S.E. and West, G. (eds) (2009), *Complexity Perspectives in Innovation and Social Change*, Berlin: Springer.

Laursen, K. and Salter, A. (2006), 'Open for innovation: The role of openness in explaining innovation performance among UK manufacturing firms', *Strategic Management Journal*, **24**, 131–50.

Lazonick, W. (1993), 'Learning and the dynamics of international competitive advantage', in R. Thomson (ed.), *Learning and Technological Change*, New York: St. Martin's, pp. 172–97.

Lazonick, W. (2007), 'Varieties of capitalism and innovative enterprise', *Comparative Social Research*, **24**, 21–69.

Le Bas, C. and Scellato, G. (2014), 'Firm innovation persistence: A fresh look at the framework of analysis', *Economics of Innovation and New Technology*, **23** (5–6), 423–46.

Lee, K.B. and Wong, V. (2011), 'Identifying the moderating influences of external environments on new product development process', *Technovation*, **31**, 598–612.

Leibenstein, H. (1976), *Beyond Economic Man: A New Foundation for Microeconomics*, Cambridge, MA: Harvard University Press.

Levinthal, D.A. (1997), 'Adaptation on rugged landscapes', *Management Science*, **43** (7), 934–50.

Levit, G., Hossfeld, U. and Witt, U. (2011), 'Can Darwinism be "generalized" and of what use would this be?', *Journal of Evolutionary Economics*, **21**, 545–62.

Lhuillery, S. (2011), 'Absorptive capacity, efficiency effect and competitors spillovers', *Journal of Evolutionary Economics*, **21**, 649–63.

Li-Ying, J., Wang, Y. and Salomo, S. (2013), 'An inquiry on dimensions of external technology search and their influence on technological innovation: Evidence from Chinese firms', *R&D Management*, **44**, 53–74.

Lin, C., Wu, Y.-J., Chang, C., Wang, W. and Lee, C.-Y. (2012), 'The alliance innovation performance of R&D alliances: The absorptive capacity perspective', *Technovation* **32**, 282–92.

Link, A.N. (1980), 'Firm size and efficient entrepreneurial activity: A reformulation of the Schumpeter hypothesis', *Journal of Political Economy*, **88**, 771–82.

Link, A. and Siegel, D. (2007), *Innovation, Entrepreneurship, and Technological Change*, Oxford: Oxford University Press.

Loasby, B.J. (2010), 'Capabilities and strategy: Problems and prospects', *Industrial and Corporate Change*, **19**, 1301–16.

Lokshin, B., Belderbos, R. and Carree, M. (2008), 'The productivity effects of internal and external R&D: Evidence from a dynamic panel data model', *Oxford Bulletin of Economics and Statistics*, **70** (3), 399–413.

Lööf, H. and Heshmati, A. (2002), 'Knowledge capital and performance heterogeneity: A firm-level innovation study', *International Journal of Production Economics*, **76** (1), 61–85.

López, R.A. (2005), 'Trade and growth: reconciling the macroeconomic and microeconomic evidence', *Journal of Economic Surveys*, **19**, 623–48.

Lopez-Garcia, P. and Puente, S. (2012), 'What makes a high growth firm? A dynamic probit analysis using Spanish firm-level data', *Small Business Economics*, **39**, 1029–41.

Lorentz, A., Ciarli, T., Savona, M. and Valente, M. (2016), 'The effect of demand driven structural transformation on growth and technological change', *Journal of Evolutionary Economics*, **26**, 219–46.

Love, J.H. and Roper, S. (2009), 'Organizing the innovation process: Complementarities in innovation networking', *Industry and Innovation*, **16**, 273–90.

Lucas, R.E. (1998), 'On the mechanisms of economic development', *Journal of Monetary Economics*, **22**, 3–42.

Lucas, R.E. (2008), 'Ideas and growth', *Economica*, **76**, 1–19.

Lundvall, B. (1988), 'Innovation as an interactive process: From user–producer interaction to the national system of innovation', in G. Dosi et al. (eds), *Technical Change and Economic Theory*, London: Frances Pinter, pp. 349–69.

Malerba, F., Orsenigo, L. and Petretto, P. (1997), 'Persistence of innovative activities sectoral patters of innovation and international technological specialization', *International Journal of Industrial Organization*, **15**, 801–26.

Malerba, F., Nelson, R.R., Orsenigo, L. and Winter, S.G. (2001), 'History-friendly models: An overview of the case of the computer industry', *Journal of Artificial Societies and Social Simulation*, **4** (3), http://jasss.soc.surrey.ac.uk/4/3/6.html.

Máñez, J., Rochina-Barrachina M. and Sanchis-Llopis, A. (2014), 'The determinants of R&D persistence in SMEs', *Small Business Economics*, **44** (3) 505–28.

Marrocu, E., Paci, R. and Pontis, M. (2012), 'Intangible capital and firms' productivity', *Industrial and Corporate Change*, **21** (2), 377–402.

Mansfield, E., Schwartz, M. and Wagner, S. (1981), 'Imitation costs and patents: An empirical study', *Economic Journal*, **91** (364), 907–18.

March, J.C. (1988), 'Bounded rationality ambiguity and the engineering of choice', in D.E. Bell, H. Raiffa and A. Tversky (eds), *Decision Making: Descriptive, Normative, and Prescriptive Interactions*, Cambridge: Cambridge University Press.

March, J.C. (1991), 'Exploration and exploitation in organizing learning', *Organization Science*, **2**, 71–87.

March, J.C. and Simon, H.A. (1958), *Organizations*, New York: Wiley.

Marrocu, E., Paci, R.and Pontis, M. (2012), 'Intangible capital and firms' productivity', *Industrial and Corporate Change*, **21** (2), 377–402.

Marshall, A. (1920 [1890]), *Principles of Economics*, 8th edition, London: Macmillan.

Martin, R. and Sunley, P. (2012), 'Forms of emergence and the evolution of economic landscapes', *Journal of Economic Behavior & Organization*, **82**, 338–51.

McEvily, S.K. and Chakravarthy, B. (2002), 'The persistence of knowledge-based advantage: an empirical test for product performance and technological knowledge', *Strategic Management Journal*, **23** (4), 285–305.

McGahan, A.M. and Silverman, B.S. (2006), 'Profiting from technological innovation by others: The effect of competitor patenting on firm value', *Research Policy*, **35** (8) 1222–42.

Merino, F. and Rodríguez, D. (2007), 'Business services outsourcing by manufacturing firms', *Industrial and Corporate Change*, **16** (6), 1147–73.

Metcalfe, J.S. (1998), *Evolutionary Economics and Creative Destruction*, London: Routledge.

Metcalfe, J.S. (2002), 'Knowledge of growth and the growth of knowledge', *Journal of Evolutionary Economics*, **12**, 3–16.

Metcalfe, J.S. (2007a), 'The broken thread: Marshall, Schumpeter and Hayek on the evolution of capitalism', ESRC Centre for Research on Innovation and Competition, University of Manchester.

Metcalfe, J.S. (2007b), 'Alfred Marshall's Mecca: Reconciling the theories of value and development', *Economic Record*, **83**, S1–22.

Metcalfe, J.S. (2009a), 'On Marshallian evolutionary dynamics, entry and exit', in J. Vint, J.S. Metcalfe, H.D. Kurz, N. Salvadori and P.A. Samuelson (eds), *Economic Theory and Economic Thought: Essays in Honour of Ian Steedman*, London: Routledge, pp. 350–73.

Metcalfe, J.S. (2009b), 'Marshall and Schumpeter: Evolution and the institutions of capitalism', in A. Pyka, U. Cantner, A. Greiner and T. Kuhn (eds), *Recent Advances in Neo-Schumpeterian Economics: Essays in Honour of Horst Hanusch*, Cheltenham, UK and Northampton, MA, USA: Edward Elgar Publishing, pp. 53–77.

Metcalfe, J.S. (2013), 'Management and the representative firm: The modern significance of the Marshall's evolutionary economics', *Economics of Innovation and New Technology*, **22**, 222–37.

Miller, J.H. and Page, S.E. (2007), *Complex Adaptive Systems*, Princeton, NJ: Princeton University Press.

Mohnen, P. and Hall, B.H. (2013), 'Innovation and productivity: An update', *Eurasian Business Review*, **3** (1), 47–65.

Mokyr, J. (1990a), 'Punctuated equilibria and technological progress', *American Economic Review*, **80** (2), 350–54.

Mokyr, J. (1990b), *The Lever of Riches*, New York and Oxford: Oxford University Press.

Mortara, L. and Minshall, T. (2011), 'How do large multinational companies implement open innovation?', *Technovation*, **31**, 586–97.

Mueller, M. and Pyka, A. (2016), 'Economic behaviour and agent-based modelling', in R. Frantz, S. Chen, S.-H., K. Dopfer, F. Heukelom and S. Mousavi (eds), *Routledge Handbook of Behavioral Economics*, London: Routledge, pp. 405–15.

Napoletano, M., Dosi, G., Fagiolo, G. and Roventini, A. (2012), 'Wage formation investment behavior and growth regimes: An agent based analysis', *Revue de l'OFCE*, **124**, 235–61.

Nelson, R.R. (1959), 'The simple economics of basic scientific research', *Journal of Political Economy*, **67**, 297–306.

Nelson, R.R. (1982), 'The role of knowledge in R&D efficiency', *Quarterly Journal of Economics*, **97**, 453–70.

Nelson, R.R. and Winter, S.G. (1977), 'In search of useful theory of innovation', *Research Policy*, **6** (1), 36–76.

Nelson, R.R. and Winter, S.G. (1982), *An Evolutionary Theory of Economic Change*, Cambridge, MA: Belknap Press of Harvard University Press.

Nelson, R.R., Winter, S.G. and Schuette, H.L. (1976), 'Technical change in an evolutionary model', *Quarterly Journal of Economics*, **90**, 90–118.

Nesta, L. (2008), 'Knowledge and productivity in the world's largest manufacturing corporations', *Journal of Economic Behavior & Organization*, **67**, 886–902.

Nesta, L. and Dibiaggio, L. (2003), 'Technology strategy and knowledge dynamics: The case of biotech', *Industry and Innovation*, **10** (3), 331–49.

Nesta, L. and Saviotti, P.P. (2005), 'Coherence of the knowledge base and the firm's innovative performance: Evidence from the pharmaceutical industry', *Journal of Industrial Economics*, **53**, 123–42.

Neter, J., Wasserman, W. and Kutner, M.H. (1990), *Applied Linear Statistical Models*, Burr Ridge, IL: Irwin.

Newell, R.G., Jaffe, A.B. and Stavins, R.N. (1999), 'The induced innovation hypothesis and energy-saving technological change', *Quarterly Journal of Economics*, **114**, 941–75.

Nooteboom, B. (2000), *Learning and Innovation in Organizations and Economies*, Oxford: Oxford University Press.

Ó hUallacháin, B. and Leslie, T.F. (2007), 'Rethinking the regional knowledge production function', *Journal of Economic Geography*, **7**, 737–52.

O'Regan, N. and Kling, G, (2011), 'Technology outsourcing in manufacturing small- and medium-sized firms: another competitive resource?', *R&D Management*, **41**, 92–105.

Ostrom, E. (2010), 'Beyond markets and states: Polycentric governance of complex economic systems', *American Economic Review*, **100**, 641–72.

Ostrom, E. and Hess, C. (eds) (2006), *Understanding Knowledge As a Commons: From Theory to Practice*, Cambridge, MA: MIT Press.

Ozman, M. (2009), 'Inter-firm networks and innovation: A survey of literature', *Economics of Innovation and New Technology*, **18**, 39–67.

Page, S.E. (2011), *Diversity and Complexity*, Princeton, NJ: Princeton University Press.

Pakes, A. and Griliches, Z. (1984), 'Patents and R&D at the firm level: A first look', in Z. Griliches (ed.), *R&D and Productivity: The Econometric Evidence*, Chicago: University of Chicago Press, pp. 55–72.

Parisi, M.L., Schiantarelli, F. and Sembenelli, A. (2006), 'Productivity innovation and R&D: Micro evidence for Italy', *European Economic Review*, **50**, 2037–61.

Parker, S., Storey, D. and van Witteloostuijn, A. (2010), 'What happens to gazelles? The importance of dynamic management strategy', *Small Business Economics*, **35**, 203–26.

Patrucco, P. (2008), 'The economics of collective knowledge and technological communication', *Journal of Technology Transfer*, **33**, 579–99.

Patrucco, P. (2009), 'Collective knowledge production costs and the dynamics of technological systems', *Economics of Innovation and New Technology*, *18*, 295–310.

Penrose, E. (1952), 'Biological analogies in the theory of the firm', *American Economic Review*, **42** (5), 804–19.

Penrose, E. (1959), *The Theory of the Growth of the Firm*, Oxford: Blackwell.

Peters, B. (2009), 'Persistence of innovation: Stylized facts and panel data evidence', *Journal of Technology Transfer*, **36**, 226–43.

Phelps, E.S. (1966), 'Models of technological progress and the golden rule of research', *Review of Economic Studies*, **33**, 133–45.

Porter, M.E. and van der Linde, C. (1995), 'Toward a new conception of the environment–competitiveness relationship', *Journal of Economic Perspectives*, **9**, (4), 97–118.

Pyka, A. and Werker, C. (2009), 'The methodology of simulation models: Chances and risks', *Journal of Artificial Societies and Social Simulation*, **12**, 1–4, http://jasss.soc.surrey.ac.uk/12/4/1.html.

Pyka, A. and Fagiolo, G. (2007), 'Agent-based modelling: a methodology for neo-Schumpeterian economics', in H. Hanusch and A. Pyka (eds), *Elgar Companion to Neo-Schumpeterian Economics*, Cheltenham, UK and Northampton, MA, USA: Edward Elgar Publishing, pp. 467–502.

Quatraro, F. (2009), 'Innovation, structural change and productivity growth: Evidence from Italian regions 1980–2003', *Cambridge Journal of Economics*, **33**, 1001–22.

Quatraro, F. (2010), 'Knowledge coherence variety and productivity growth: Manufacturing evidence from Italian regions', *Research Policy*, **39**, 1289–302.

Quatraro, F. (2012), *The Economics of Structural Change in Knowledge*, London: Routledge.

Rahko, J. (2014), 'Market value of R&D, patents, and organizational capital: Finnish evidence', *Economics of Innovation and New Technology*, **23** (4), 353–77.

Ravix, J.L. (2012), 'Alfred Marshall and the Marshallian theory of the firm', in M. Dietrich and J. Krafft (eds), *Handbook on the Economics and the Theory of the Firm*, Cheltenham, UK and Northampton, MA, USA: Edward Elgar Publishing, pp. 49–54.

Raymond, W., Mairesse, J., Mohnen, P. and Palm, F. (2013), 'Dynamic models of R&D, innovation and productivity: Panel data evidence for Dutch and French manufacturing', NBER Working Paper 19074.

Reichman, J. (2000), 'Of green tulips and legal kudzu: Repackaging rights in subpatentable invention', *Vanderbilt Law Review*, **53**, 17–43.

Rigby, D.L. (2015), 'Technological relatedness and knowledge space: Entry and exit of US cities from patent classes', *Regional Studies*, **49** (11), 1922–37.

Rivera-Batiz, L.A. and Romer, P.M. (1991), 'Economic integration and economic growth', *Quarterly Journal of Economics*, **106**, 531–56.

Roberts, V., Yoguel, G. and Lerena, O. (2017), 'The ontology of complexity and the neo-Schumpeterian evolutionary theory of economic change', *Journal of Evolutionary Economics*, **27** (4) 761–94.

Romer, P.M. (1990), 'Endogenous technological change', *Journal of Political Economy*, **98**, S71–102.

Roper, S. and Hewitt-Dundas, N. (2008), 'Innovation persistence: Survey and case-study evidence', *Research Policy*, **37**, 149–62.

Rosser, J.B. (1999), 'On the complexities of complex economic dynamics', *Journal of Economic Perspectives*, **13**, 169–92.

Rosser, J.B. (ed.) (2004), *Complexity in Economics: Methodology Interacting Agents and Microeconomic Models*, Cheltenham, UK and Northampton, MA, USA: Edward Elgar Publishing.

Rothaermel, F.T. and Hess, A.M. (2007), 'Building dynamic capabilities', *Organization Science*, **18** (6), 898?921.

Ruckman, K. (2008), 'Externally sourcing research through acquisition: Should it supplement or substitute for internal research?', *Industry and Innovation*, **15**, 627–45.

Ruttan, V.W. (1997), 'Induced innovation evolutionary theory and path dependence', *Economic Journal*, **107** (444), 1520–29.

Safarzyńska, K. and van den Bergh, J.C.J.M. (2010), 'Evolutionary models in economics: A survey of methods and building blocks', *Journal of Evolutionary Economics*, **20** (3), 329–73.

Santarelli, E. and Vivarelli, M. (2007), 'Entrepreneurship and the process of firms' entry, survival and growth', *Industrial and Corporate Change*, **16** (3), 455–88.

Saviotti, P.P. (2004), 'Considerations about the production and utilization of knowledge', *Journal of Institutional and Theoretical Economics*, **160**, 100–121.

Saviotti, P.P. (2007), 'On the dynamics of generation and utilisation of knowledge: The local character of knowledge', *Structural Change and Economic Dynamics*, **18**, 387–408.

Saviotti, P.P. and Pyka, A. (2008), 'Micro and macro dynamics: Industry life cycles, inter-sector coordination and aggregate growth', *Journal of Evolutionary Economics*, **18**, 167–82.

Schankerman, M. and Pakes, A. (1986), 'Estimates of the value of patent rights in European countries during the post-1950 period', *Economic Journal*, **96** (384), 1052–76.

Scherer, F.M (1986), *Innovation and Growth: Schumpeterian Perspectives*, Cambridge, MA: MIT Press.

Schumpeter, J.A. (1911–34), *The Theory of Economic Development*, Cambridge, MA: Harvard University Press.

Schumpeter, J.A. (1928), 'The instability of capitalism', *Economic Journal*, **38**, 361–86.

Schumpeter J.A. (1939), *Business Cycles: A Theoretical, Historical and Statistical Analysis of the Capitalist Process*, New York: McGraw-Hill.

Schumpeter, J.A. (1941), 'Alfred Marshall's Principles: A semi-centennial appraisal', *American Economic Review*, **31** (2), 236–48.

Schumpeter, J.A. (1942), *Capitalism, Socialism, and Democracy*, New York: Harper & Brothers.

Schumpeter, J.A. (1947), 'The creative response in economic history', *Journal of Economic History*, **7**, 149–59.

Scitovsky, T. (1954), 'Two concepts of external economies', *Journal of Political Economy*, **62**, 143–51.

Silva, E. and Teixeira, A.C. (2009), 'Surveying structural change: Seminal contributions and a bibliometric account', *Structural Change and Economic Dynamics*, **19**, 273–300.

Simon, H.A. (1947), *Administrative Behavior: A Study of Decision-Making Processes in Administrative Organization*, London: Macmillan.

Simon, H.A. (1979), 'Rational decision making in business organizations', *American Economic Review*, **69** (4), 493–513.

Simon, H.A. (1982), *Metaphors of Bounded Rationality: Behavioral Economics and Business Organization*, Cambridge, MA: MIT Press.

Smit, M.J., Abreu, M.A. and de Groot, H.L. (2015), 'Micro-evidence on the determinants of innovation in the Netherlands: The relative importance of absorptive capacity and agglomeration externalities', *Papers in Regional Science*, **94** (2), 249–72.

Sorenson, O., Rivkin, J.W. and Fleming, L. (2006), 'Complexity, networks and knowledge flow', *Research Policy*, **35**, 994–1017.

Stock, J.H. and Yogo, M. (2005), 'Testing for weak instruments in linear IV regression', in D.W.K. Andrews and J.H. Stock (eds), *Identification and Inference for Econometric Models: Essays in Honor of Thomas Rothenberg*, Cambridge: Cambridge University Press, pp. 80–108.

Strumsky, D., Lobo, J. and Van der Leeuw, S. (2012), 'Using patent technology codes to study technological change', *Economics of Innovation and New Technology*, **21**, 267–86.

Suarez, D. (2014), 'Persistence of innovation in unstable environments: Continuity and change in the firm's innovative behavior', *Research Policy*, **43**, 726–36.

Syverson, C. (2011), 'What determines productivity?', *Journal of Economic Literature*, **49** (2), 326–65.

Teece, D. (2007), 'Explicating dynamic capabilities: The nature and microfoundations of (sustainable) enterprise performance', *Strategic Management Journal*, **28** (13), 1319–50.

Teece, D. and Pisano, G. (1994), 'The dynamic capabilities of firms: An introduction', *Industrial and Corporate Change*, **3**, 537–55.

Terna, P. (2009), 'The epidemics of innovation: Playing around with an agent-based model', *Economics of Innovation and New Technology*, **18**, 707–28.

Theyel, N. (2013), 'Extending open innovation throughout the value chain by small and medium-sized manufacturers', *International Small Business Journal*, **31** (3), 256–74.

Thoma, G., Torrisi, S., Gambardella, A., Guellec, D., Hall, B.H. and Haroff, D. (2010), 'Harmonizing and combining large datasets: An application to firm-level patent and accounting data', NBER Working Paper 15851.

Triguero, A. and Corcoles, D. (2013), 'Understanding innovation: An analysis of persistence for Spanish manufacturing firms', *Research Policy*, **42**, 340–52.

Triguero, A., Corcoles, D. and Cuerva, M.C. (2014), 'Persistence of innovation and firm's growth: Evidence from a panel of SME and large Spanish manufacturing firms', *Small Business Economics*, **43** (4), 787–804.

Van Zeebroeck, N. (2011), 'The puzzle of patent value indicators', *Economics of Innovation and New Technology*, **20**, 33–62.

Van Zeebroeck, N. and van Pottelsberghe, B. (2011), 'The vulnerability of patent value determinants', *Economics of Innovation and New Technology*, **20**, 283–308.

Vandekerckhove, J. and De Bondt, R. (2008), 'Asymmetric spillovers and investments in research and development of leaders and followers', *Economics of Innovation and New Technology*, **17**, 417–33.

Veblen, T. (1898), 'Why is economics not an evolutionary science?', *Quarterly Journal of Economics*, **2** (4), 373–97.

Vergne, J.P. and Durand, R. (2011), 'The path of most persistence: an evolutionary perspective on path dependence and dynamic capabilities', *Organization Studies*, **32**, 365–82.

Verona, G. and Ravasi, D. (2003), 'Unbundling dynamic capabilities: An exploratory study of continuous product innovation', *Industrial and Corporate Change*, **12** (3), 577–607.

Veugelers, R. and Cassiman, B. (1999), 'Make and buy in innovation strategies: Evidence from Belgian manufacturing firms', *Research Policy*, **28**, 63–80.

Von Hippel, E. (1988), *The Sources of Innovation*, Oxford: Oxford University Press.

Von Hippel, E. (1998), 'Economics of product development by users: The impact of sticky local information', *Management Science*, **44**, 629–44.

Voudouris, I., Lioukas, S., Iatrelli, M. and Caloghirou, Y. (2012), 'Effectiveness of technology investment: Impact of internal technological capability, networking and investment's strategic importance', *Technovation*, **32** (6), 400–414.

Weitzman, M.L. (1996), 'Hybridizing growth theory', *American Economic Review*, **86**, 207–12.

Weitzman, M.L. (1998), 'Recombinant growth', *Quarterly Journal of Economics*, **113**, 331–60.

Windrum, P. and Birchenhall, C. (2005), 'Structural change in the presence of network externalities: A co-evolutionary model of technological successions', *Journal of Evolutionary Economics*, **15**, 123–48.

Winter, S.G. (1987), 'Knowledge and competence as strategic assets', in D.J. Teece (ed.), *The Competitive Challenge: Strategies for Industrial Innovation and Renewal*, Cambridge, MA: Ballinger, pp. 159–84.

Winter, S.G., Kaniovski, Y.M. and Dosi, G. (2000), 'Modeling industrial dynamics with innovative entrants', *Structural Change and Economic Dynamics*, **11** (3), 255–93.

Wooldridge, J. (2005), 'Simple solutions to the initial conditions problem in dynamic, nonlinear panel data models with unobserved heterogeneity', *Journal of Applied Econometrics*, **20**, 39–54.

Yildizoglu, M. (2002), 'Competing R&D strategies in an evolutionary industry model', *Computational Economics*, **19**, 51–65.

Youn, H., Bettencourt, L.M.A., Strumsky, D. and Lobo, J. (2015), 'Invention as a combinatorial process: Evidence from US patents', *Journal of the Royal Society, Interface*, **12** (106), DOI: 10.1098/rsif.2015.0272.

Zhang, J. (2003), 'Growing Silicon Valley on a landscape', *Journal of Evolutionary Economics*, **13**, 529–48.

Index